Robert Willsher Weekes

The Design of Alternate-Current Transformers

Robert Willsher Weekes

The Design of Alternate-Current Transformers

ISBN/EAN: 9783337185916

Printed in Europe, USA, Canada, Australia, Japan

Cover: Foto ©berggeist007 / pixelio.de

More available books at **www.hansebooks.com**

ALTERNATE-CURRENT TRANSFORMERS.

THE DESIGN

OF

ALTERNATE-CURRENT

TRANSFORMERS.

BY

R. W. WEEKES, WHIT.SCH., ASSOC.M.INST.C.E.

ILLUSTRATED.

BIGGS AND CO., 139-140, SALISBURY COURT, LONDON, E.C

1893.

PREFACE.

THE invention of the alternate-current transformer some years ago led to the introduction of the system of high-tension distribution of electric energy for lighting purposes. The saving in the first cost of the copper mains being the chief advantage of the system in the eyes of the promoters, the question of the efficiency of this method of supply, except at full load, was not considered. Now, experience has shown that the cost of the annual supply depends largely on the efficiency of the transformers used, and that the loss at light loads is the most important factor in determining this cost.

Hence the design of alternate-current transformers needs more careful attention as the demand for higher efficiency arises.

I have endeavoured in the following pages to describe as simply as possible the principles involved in the construction of transformers on economic lines, both as regards first cost and power wasted when working. To the formulæ deduced from these principles I have added several others which are useful when determining quickly the dimensions of the various circuits. To enable more to follow the reasoning, higher mathematics have been avoided, although this omission has necessitated the acceptance of certain well-known formulæ without proof.

The examples worked out and illustrated show the general procedure of designing transformers of different types to fulfil definite conditions.

R. W. WEEKES.

CONTENTS.

	PAGE.
Brown-Boveri Transformer	46
Circuit, Iron	74, 77
Circuits, Perimeter of	21, 47, 49, 69
Circuits, Size, Copper	23, 50, 52, 53, 71
Conclusions	93
Condensers, Use with Transformers	80
Cooling Surface	22, 39, 57, 73
Copper Circuit, First Design	30, 36
Copper in Circuits	23, 50, 52, 53, 71
Copper Loss	13, 38, 56, 67, 73
Core...	43, 59, 61, 65, 71, 75, 77
Design, Long Shell Type, Square Core	29
Design, Mordey Type	47
Design, Third	65
Designing, Routine in	25
Details, First Design	37
Details, Second or Mordey Design ...	56
Details, Third Design	72
Early Designs, Faults in	10
Efficiency	13, 28, 84
Efficiency and Frequency	84
Elwell-Parker Transformer	62
E.M.F. in Coil	15
Faults in Early Designs	10
Ferranti Transformer	71
First Design	29
Fleming's Table, Transformers Tested	11
Foucault Currents	19, 21, 69
Frequency	25, 29, 84, 85
Frequency and Efficiency	84
Hedgehog Transformer	76
Hysteresis ...	19, 20, 87, 91

CONTENTS.

	PAGE
Induction	19, 86
Iron Cores	43, 59
Iron Loss	12, 20, 35, 49, 56, 66, 73, 79, 80, 89, 90
Iron, Quality of	15
Joints, Faced	45
Large v. Small Transformers	81
Leakage, Magnetic	24, 64
Magnetic Leakage	24, 64, 81
Mordey Type Transformer	47
Oerlikon Transformer	44
Permeability	19, 25, 36
Power Factor	20, 58
Principles involved	14
Routine in Designing	25
Sine Law	15
Stampings	48
Steinmetz on Hysteresis	20
Swinburne's Transformer	76
Tests, Fleming's	11
Transformer Formula for E.M.F.	16
Transformer Loads	18
TRANSFORMERS—	
Brown-Boveri	46
for Different Methods of Supply	94
Elementary	14
Elwell-Parker	62
Ferranti	71
Large v. Small	81
Law of	15
Oerlikon	44
Table of Tests	11
Westinghouse	61
Value of μ	19
Voltage	16, 17
Westinghouse Transformer	61
Windings, Determination, First Design	30
Windings, Mordey Type	54
Wire, First Design	33, 34, 39

THE DESIGN

OF

ALTERNATE CURRENT TRANSFORMERS.

The reliable information available on the subject of alternate-current transformers has until recently been very scanty. When the high tension system of distribution, with transformers in parallel, was introduced some eight years ago, the good points of the transformers as then made, were much exaggerated. It was generally stated that the efficiency at all loads was very high, and that the current taken by the primary when no lamps were on the secondary circuit was so small as to be almost neligible. The last statement may be excused to some extent when it is remembered that small current measurements could not be made so readily then. Still, the engineers in charge of central stations have found that the small currents required to energise each transformer soon become a serious item in the cost of production. In consequence of their experience they now specify what the various transformers should be capable of doing, and by raising the standard of their requirements induce the makers to make further improvements.

The faults existing in the early designs of transformers may be briefly summarised as follows: Bad regulation, that is, the pressure at the secondary terminals varied considerably with the load. Hence, when the primary circuit was supplied at a constant difference of potential, the consumer found that the voltage fell as he increased the number of the lamps burning. This fault was partly overcome by increasing the primary voltage slightly, when the load came on, but this does not remove the fault unless all consumers take the full load at the same hours. Otherwise, in some installations, a few lamps might be seriously overrun by the rise in pressure.

Another more important fault, was the comparatively large current taken by the primary when there was no load on the secondary. This is due in a large measure to the loss by hysteresis and Foucault currents in the iron core, which in some of the early transformers amounted to from 10 to 20 per cent. of the total output. In addition to these two principal failings, there were several others in a large measure resulting from them, such as the high temperature of the transformer even when not loaded, faulty insulation between the two windings, etc.,

The various makers have since that time done much to reduce these faults to a minimum, but the methods used to this end and the general practice of the design of transformers is still known to comparatively few. This is partly due to the tendency amongst many writers on the subjects to treat

Alternate Current Transformers.

LIST OF TRANSFORMERS TESTED BY DR. FLEMING.

$\frac{\frown}{\frown} = 83.$

Transformers.	Maximum output in watts from secondary.	Magnetising current in amperes.	Primary volts.	Power absorbed in watts at no load.	Apparent watts at no load.	Power factor.	Iron loss in per cent. of full load.	Magnetising current in per cent. of full current.	Total drop at full load in volts.	Copper drop in volts.	Leakage drop in volts.
Ferranti (1885 type)	1,875	·18	2,416	288	432	·66	15·4	23	—	—	—
,, ,,	3,750	·337	2,400	540	808	·68	14·6	21·6	1·6	1·9	—
,, ,,	7,500	·25	2,435	444	600	·74	5·9	8·1	—	—	—
,, ,,	11,250	·34	2,447	578	816	·70	5·15	7·4	—	—	—
,, ,,	15,000	·57	2,389	1,019	1,368	·75	6·8	9·0	—	—	—
,, ,, (1885 rewound)	3,750	·11	2,400	233	264	·88	6·2	7·0	2·4	2·15	·25
,, ,, (1892 type)	7,500	·075	2,400	138	180	·77	1·84	2·4	—	—	·65
,, ,,	11,250	·076	2,400	148	182	·81	1·31	1·61	3·4	2·75	·65
,, ,,	15,000	·112	2,400	228	269	·85	1·52	1·79	2·1	1·65	·45
,, ,, (1892 rewound)	11,250	·103	2,400	228	247	·92	2·02	2·2	2·2	1·78	·42
Swinburne Hedgehog	3,000	·74	2,400	112	1,775	·063	3·73	59·0	3·2	2·23	·97
,, ,,	6,000	1·2160	2,400	165	2,920	·05	2·75	47·5	—	—	—
Westinghouse	6,500	·05	2,400	95	120	·79	1·46	1·85	2·4	1·38	1·02
Mordey Brush	6,000	·076	2,400	140	182	·77	2·33	3·05	1·8	1·75	—
,, ,,	750	·0317	2,392	61·5	76	·81	8·2	10·2	—	—	—
Thomson-Houston	4,500	·083	2,400	108	199	·54	2·4	4·42	3·3	2·47	·83
Kapp	4,000	·145	2,400	152	348	·61	3·8	8·7	1·9	1·83	—

transformer design as a field for mathematical exercise. In this way a lot of really unnecessary mathematics is often introduced which frightens the great majority of electrical engineers.

The recent paper by Dr. Fleming, read before the Institution of Electrical Engineers in November, 1892 formed the most practical contribution to the literature on the subject. The list given in the paper of the actual losses in some 18 different transformers is a rough guide as to what may be expected of the modern transformer. It must be remembered when examining the table which is given before, that the transformer-makers are rapidly improving their manufactures, and hence some of the results may be already out of date. The comparison of the Ferranti transformers made in 1892 with those of 1885 will show the vast difference between the old types and those perfected by careful study.

In addition to the columns found in Dr. Fleming's table, I have added five others, showing the percentage of magnetising current and iron loss in each case, and also the figures relating to the regulation of the various transformers which are taken from other parts of the same paper. In this form the list is of great value for reference when getting out new designs.

All the transformers in the list were wound and reduced from 2,400 volts to 100.

The importance of reducing the iron loss will be seen when it is remembered that an average private house takes from five to twenty Board of Trade units per year per 8-c.p. lamp installed. If we consider a house having 120 of such lamps fixed, it might safely be supplied by the second transformer on the list. During the year it might take and pay for 13 units per lamp, or $120 \times 13 = 1,560$ units for the whole supply. But the transformer has 14·6 per cent. of iron loss, or wastes 545 watts continuously. Taken over the 8,760 hours in the year, this amounts to $\frac{545 \times 8,760}{1,000}$ units $= 4,760$ units. That means that four times the quantity paid for has to be generated. The efficiency of distribution is then $\frac{1,560}{6,320} = 24\cdot6$ per cent., neglecting the loss in the line. Now, if the iron loss were reduced to 2 per cent, which may be reasonably expected, the annual waste in the iron is then $\frac{75 \times 8,760}{1,000} = 655$ units; so that, neglecting copper loss, the efficiency of distribution rises to 70 per cent. With other consumers, such as clubs and shopkeepers, the efficiency will be considerably over 90 per cent. if the load is steady for any length of time.

The copper loss is important at full load, but decreases as the square of the current used, and so may be neglected at light loads. Thus, if the copper loss is 2 per cent. at full load it falls to ·5 per cent. at half load, ·125 per cent. at one-quarter full load, and to ·02 per cent. at one-tenth of full load. Hence, the regulation required does more to determine the copper loss allowable than the consideration of the actual waste in the copper circuits.

Before going into the actual calculation of a

transformer it will be well to give a short description of the fundamental principles involved. The discovery by Faraday that any change of induction in an iron core surrounded by a coil of wire caused an instantaneous E.M.F. in the wire, is employed in transformers, and the E.M.F. is made practically continuous for heating and lighting purposes by an ever-changing induction in the iron. The elementary transformer, Fig. 1, consists of an

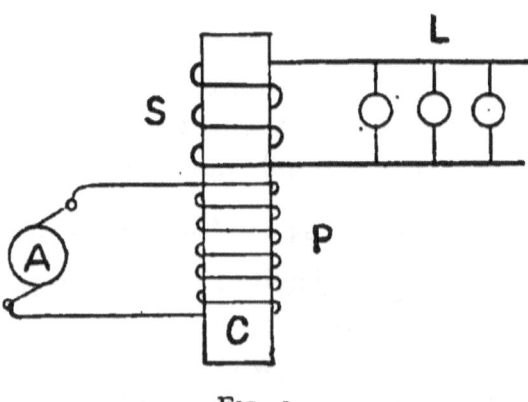

FIG. 1.

iron circuit, C, and two copper circuits, P and S, so arranged that the fluctuating current in the primary, P, causes a change of induction in the iron core, which again sets up a current in the other copper curcuit, S.

It is in determining the best relative proportions of these circuits for any given result that the art of the designer has to be applied. He was always dependent on the iron manufacturer for the quality of the iron he uses, and needs to keep a constant watch by testing the iron as it is delivered. In spite of this disturbing and ever altering element in the sequence of the designs, it is wrong to blame the iron manufacturer for all the faults which may occur, as is often done. The relative value of two designs is not altered much by a change in the quality of the iron used. Hence, the calculations made, if carefully recorded, together with the quality of iron assumed, have always a relative value, and can easily be altered to show the improvement made in the transformer if a better brand of iron can be obtained.

In the following investigations it is assumed that the E.M.F. curve of the alternator follows the simple sine law. This is strongly objected to by some writers, who advocate the method of taking the actual curves from the alternators with which the transformers have to be used. The published curves of alternators made in England, however, differ very slightly from the curve of sines, and the time wasted in hunting up curves and introducing the corrections from them in the formulæ would not be worth the slight additional accuracy obtained. The makers would not think of altering their stock patterns for so slight a difference in the working conditions as will be found between the E.M.F. curves of different types of alternators, and hence it is better in every way to work on the above assumption.

The E.M.F. generated in a coil of wire through which the induction is varied is equal to the product of the number of turns into the rate of change of the induction, all these quantities being expressed in absolute units. From this law it follows that in a transformer the volts as measured in a Cardew voltmeter are given for the primary circuit by the expression—

$$e_1 = \pi \sqrt{2}\, F\, \tau_1\, n\, 10^{-8},$$

where e_1 = the E.M.F. as defined above, in the primary;

e_2 = the E.M.F. as defined above, in the secondary;

F = the maximum total flux = $\mathfrak{B} \times a$;

a = the area of the iron in the core in square centimetres:

$\mathfrak{B}$ = the maximum induction in the core;

τ_1 = the number of turns in the primary;

τ_2 = the number of turns in the secondary;

n = the number of complete periods per second—a period being an oscillation to and fro;

10^{-8} being the coefficient which reduces the E.M.F. from absolute units to volts.

The above formulæ can be best remembered as

$$e_1 = 4\cdot 45\, F\, \tau_1\, n\, 10^{-8} \quad . \quad . \quad . \quad (1)$$

Similarly the E.M.F. in the secondary is given by

$$e_2 = 4\cdot 45\, F\, \tau_2\, n\, 10^{-8}.$$

Dividing one by the other we get

$$\frac{e_1}{e_2} = \frac{\tau_1}{\tau_2},$$

or, that the number of turns wound on each coil must be made proportional to the voltage required in the windings.

It is also seen that with a fixed frequency and number of turns, any given voltage is produced by a definite flux through the core. A mental picture of the equilibrium maintained is useful in making this clear. When the secondary circuit is open we require that the back E.M.F. in the primary shall rise to a value almost equal to the potential difference applied, otherwise a large current will pass; but a large current in the primary with the secondary open means that the core of the transformer will be strongly magnetised, and hence give a large back E.M.F. Thus, when the magnetising current has risen to such a value as to produce the flux given by the formulæ above, no further rise is possible, as it would cause the transformer to give power back to the mains because the back E.M.F. would be higher than that on the mains.

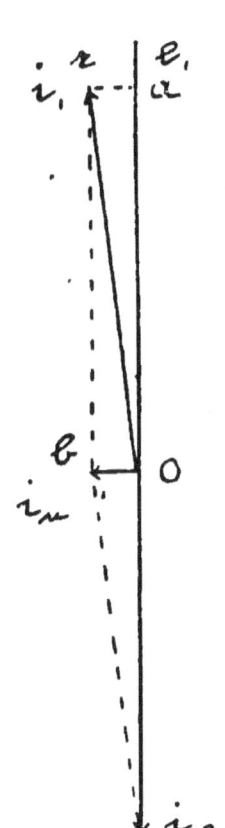

FIG. 2.

From these considerations it also follows that the induction in the core must be practically constant at all loads, because the primary circuit is supplied at constant potential. Thus while the primary current is increased by loading the secondary, the resultant magnetising current must be the same. This is explained by the fact that the primary and secondary currents practically oppose each other in their action on the core.

Fig. 2 shows the condition of affairs when the transformer is loaded.

e_1 represents in magnitude and direction the primary voltage applied;

i_1 is the primary current, which lags a little behind e_1;

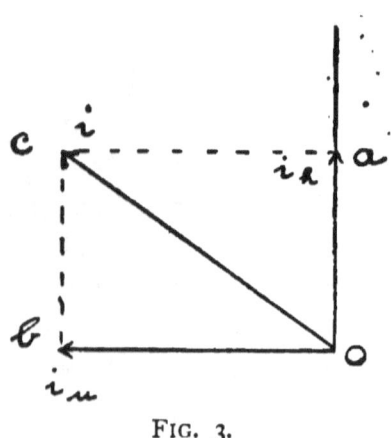

FIG. 3.

i_2 is the secondary current, which is practically opposed in phrase to e_1;

e_2, the secondary E.M.F., would be in the same line provided the external load is non-inductive.

The resultant of i_1 and i_2 obtained by the usual construction is O b, which represents the current actually required to magnetise the iron. This is here called $i\mu$, as the permeability of the iron determines its magnitude. Now i_1 can be split up into two components, O a, O b by projection; of these two, O a represents the power given to the primary, and when multiplied by e_1, gives the actual power in watts and the O b represents the current called $i\mu$ above.

Now when the secondary circuit is open, the only load on the primary will be that due to the hysteresis and the Foucault currents in the core. So in Fig. 3, if we put up O a such that O $a \times e_1$ equals the watts wasted in the core, and then make O b the same as in Fig. 2, we get the resultant of these two currents in O c. The scale is larger than that in the last figure for sake of clearness. Hence O c represents the current actually taken by the transformer when the secondary is open. These two components can be calculated if the permeability of the iron and its loss by hysteresis and Foucault currents at different inductions have been obtained. The total watts lost as obtained from the curves, are divided by e_1 to get i_H.

From the value of μ for the iron as the induction used we get by first principles that

$$i\mu = \frac{\mathfrak{B} l}{\mu \tau_1 \, 1\cdot 76 *} \qquad \ldots \ldots \quad (2)$$

* For $\mu = \dfrac{\mathfrak{B}}{H}$, and $H = \dfrac{1\cdot 25 \times \text{ampere-turns}}{l}$;

$$= \frac{1\cdot 25 \times 1\cdot 41 \times i\mu}{l};$$

$$= \frac{1\cdot 76 \times i\mu}{l};$$

$$\therefore \mu = \frac{\mathfrak{B} l}{1\cdot 76 \, i \, \mu \, \tau_1};$$

and $$i\mu = \frac{\mathfrak{B} l}{\mu \, \tau_1 \, 1\cdot 76}.$$

where $l=$ the mean length of the path taken by the induction in centimetres;

$\mathfrak{B} =$ the maximum induction in C.G.S. units;

$\mu =$ the permeability of the iron at that induction;

$\tau_1 =$ the number of turns on the primary coil.

Having thus obtained $i\mu$, we can get i the no load current because as $i_\text{H} \times i\mu$ thus form the two sides of a right angled triangle, their resultant i is got by taking the square root of the sum of their square or

$$OC^2 = CB^2 + OB^2$$
$$i^2 = i_\text{H}^2 + i\mu^2$$
$$i = \sqrt{i_\text{H}^2 + i\mu^2}.$$

The ratio of the current taken by the transformer at no load to the current actually required for the hysteresis has been called the "power factor" by Dr. Fleming, and this gives us the factor by which the apparent watts have to be multiplied to get the true watts wasted. For good closed circuit transformers this factor varies from ·60 up to nearly ·90.

As regards the curves of iron losses. These should be carefully obtained for each sample of iron used in the transformers. It is possible to subdivide the loss into two—that due to eddy currents in the iron and that due to hysteresis pure and simple. Mr. C. P. Steinmetz has worked continuously at this subject for some years, and concludes that the hysteresis losses vary as the frequency and as the 1·6th power of the induction, or $H = a\, n\, \mathfrak{B}^{1\cdot6}$; while the Foucault currents vary as $(n\, \mathfrak{B})^2$, or $F = b\, (n\, \mathfrak{B})^2$, where a and b are constants.

The simplest way is to calculate the Foucault current loss, which if the core is well subdivided, is not a large percentage of the whole, and then to get the hysteresis loss by subtraction. For instance, using the generally accepted formulæ,* the eddy currents in a pound of iron subjected to an induction of 4,000 at a frequency of 100 complete periods per second works out to ·08 watt if the plates are ten mils thick, and to ·18 watt if the plates are 15 mils thick. The watts lost per pound in hysteresis will vary about the region of one watt per pound, being more or less according to the quality of the iron. So that with the thin plates the eddy-current loss is under 8 per cent. of the total. It is also convenient to plot the total iron loss for a given frequency at different inductions, and such a curve will be added for reference.

The other theoretical considerations to be remembered in designing a transformer are of a more general nature, such as keeping the copper circuits of as small a perimeter as possible. The same applies to the iron circuits, as any undue length in the iron increases both the iron loss and the idle current when the transformer is unloaded. In fact, these points are so easily expressed in formulæ that there have been many attempts made to deduce practical conclusions as to the best design from these alone. The results thus obtained have not been

* For thin iron plates the watts lost per cubic centimetre at 0° F,
$$= \gamma = (t \, B \, n)^2 \, 10^{-10}$$
where t = the thickness of the plate in inches
B = the maximum induction
n = the frequency.

satisfactory as a whole, and sometimes have led to mistakes in practice.

The question of the cooling surface to be allowed is one of the most vital importance. The custom of enclosing transformers in cast-iron cases tends to increase the difficulty of preventing the temperature becoming too high. The heat generated in the transformer has to pass first from it into the surrounding air inside the case, then from this layer of air into the case, and finally from the case into the surrounding medium. This triple transference is greatly assisted if oil is used to fill in the case. The oil acts by conduction and convection in carrying the heat to the case, and thus materially reduces the final temperature of the transformer. The oil is primarily said to be used for insulation purposes, but its action as a cooling agent is at low voltages the more valuable of its properties.

The actual design of the sections of the two copper circuits for a given loss is a problem similar to those occurring in dynamo calculation, and presents no new difficulties.

The following formulæ, which are simply deductions from the specific resistance of copper will be found very useful in determining the sizes of the copper circuits:

If $s=$ the section of the wire in square inches;
$d=$ the diameter of the wire in inches;
$i=$ the maximum current in amperes, as read on a Siemens dynamometer;
$\tau=$ the number of turns in the circuit;
$\pi=$ the mean length of a turn in feet;

$l=$ the total length of conductor in feet;
$\theta=$ the drop of volts allowed at maximum current;
$r=$ the resistance in ohms;
$g=$ the weight of copper in pounds.

Then for rectangular sections

$$\left. \begin{array}{l} s = \dfrac{i\,\tau\,\pi\,9\cdot 2}{\theta}\,10^{-6} \\[6pt] r = \dfrac{\tau\,\pi\,9\cdot 2}{s}\,10^{-6} \\[6pt] g = s\,\tau\,\pi \times 3\cdot 85 \end{array} \right\} \quad \ldots \ldots (3)$$

$\tau\,\pi$ can in each case be replaced in the above by l, but it is well to record the mean diameter in each design.

With wire of circular section these formulæ become

$$\left. \begin{array}{l} d = \sqrt{\dfrac{i\,\tau\,\pi}{\theta}\,11\cdot 75}\,10^{-3} \\[6pt] r = \dfrac{\tau\,\pi\,11\cdot 75}{d^2}\,10^{-6} \\[6pt] g = d^2\,\tau\,\pi\,3\cdot 02 \end{array} \right\} \quad \ldots \ldots (4)$$

The constants 9·2 and 11·75 give the respective sizes and resistances at a temperature of about 105deg. F., which may be considered as a fair allowance for heating, and on assumption that the copper is of 100 per cent. conductivity. If a higher temperature is allowed, or inferior copper is used, these constants must be modified accordingly.

There is, however, one other important point to be considered, and that is the question of magnetic leakage. When most transformers are tested, it will be found that the drop in volts in the secondary

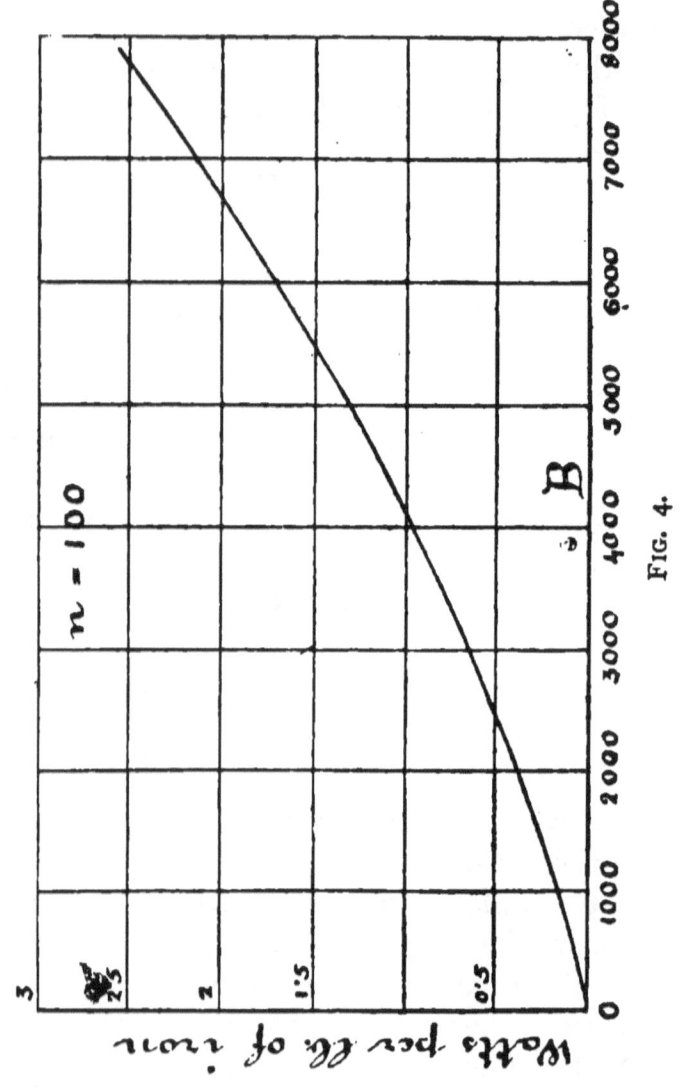

Fig. 4.

between no load and full load is more than can be accounted for by the resistances of the copper circuits. This is due to the fact that at full load the whole field produced by the primary does not pass through the secondary, but some portion of it is forced out by the action of the secondary current, and consequently we get a drop in volts. This is an important point, but it will be more readily understood after the trial designs have been given.

It will be best to assume a given quality of iron taken from practical test, and hence the two curves given have been prepared. Fig. 4 gives the watts lost per pound of iron at 100 frequency with varying induction. For slight alterations of frequency it will not cause a serious error to assume that the loss varies directly as the number of complete periods per second, as the plates for which the curve is drawn were 10 mils thick. The iron is not the best that can be obtained, but still is a fairly good sample. The other curve, Fig. 5, giving the permeability, μ, for the various inductions, was also calculated from actual results with the same iron, and will be used when working out the open-circuit currents. It will be noticed that the curve gradually rises till the induction reaches 7,000, and after that it is nearly flat. If carried farther the curve falls again, but it is not usual to work at higher inductions. With superior qualities of iron the μ curve is raised throughout its entire length, but the most noticeable feature of the change is that the permeability at low induction is very much increased.

The following is a useful routine to adopt in making designs: First fix the iron and copper losses to be allowed from the considerations of the work to be done. Then fix roughly the dimensions of the iron circuit and calculate the weight of iron. From

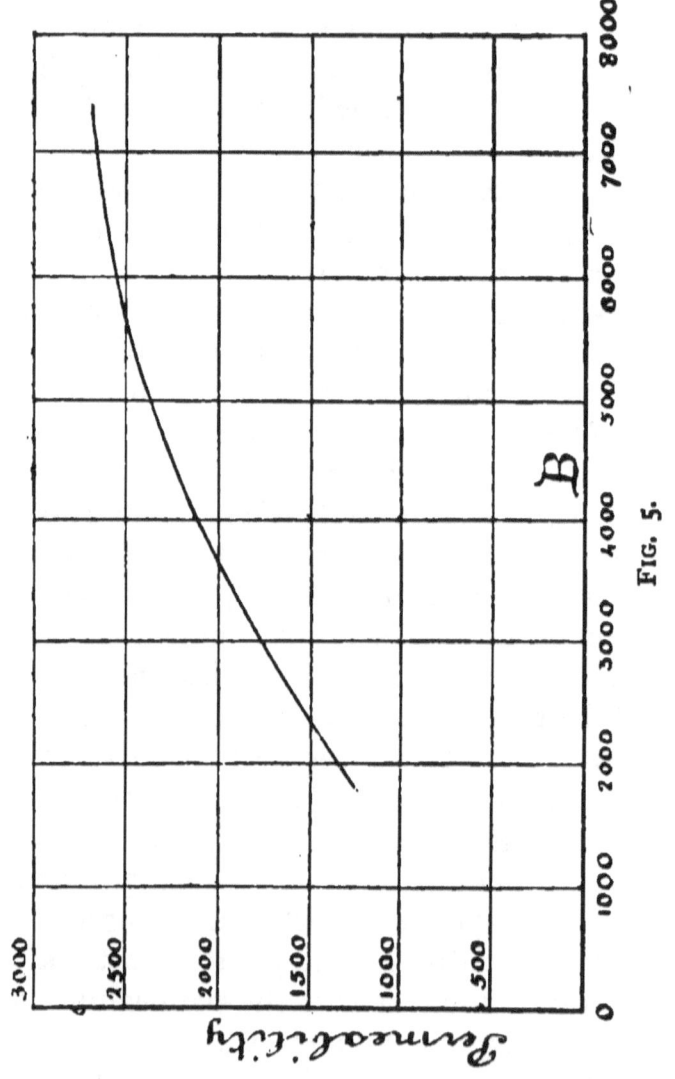

FIG. 5.

the iron loss obtain the loss per pound, and hence, from Fig. 4, the induction. This, multiplied by the cross-section of the iron, gives the total flux, and from formula (1) the number of turns required in each winding can then be found. Obtaining the mean length of turn approximately, the sections or diameters of the wires can be found, together with the weights for the assumed copper loss, from formulæ (3) and (4). The question then becomes one of mechanical design to get the winding arranged symmetrically with ample insulation. It will usually be found necessary to alter slightly the iron circuit, and very often the other details may require modification. The main elements in the design being fixed, the calculations of iron loss and magnetising current can be carefully gone through, and finally the whole data should be collected into a concise form for comparison with other designs. The watts lost per square inch cooling surface at no load and full load should be recorded, although the different types are not directly comparable in this respect. The results thus obtained are sure to suggest various alterations likely to effect improvements.

The most economical plan, however, is to make many designs for two outputs varying considerably so as to obtain the most economical forms for these, and then to get out the intermediate sizes by aid of interpolation. In this way the merits of any given type form a separate investigation, and the error of comparing a badly-proportioned transformer of one type with a well-designed transformer of another type may be avoided.

I propose to work out designs for a six-kilowatt transformer of three different types to explain the above method of working. The reader can then for himself try the effects of small alterations in the different circuits on the efficiency and cost. The following may

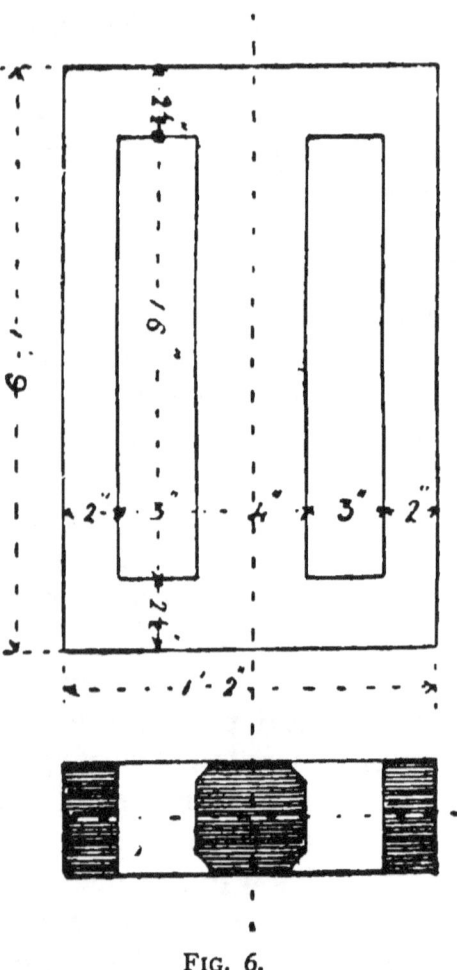

FIG. 6.

be taken as a fair efficiency to be aimed at as a first attempt: Iron loss, 150 watts = $2\frac{1}{2}$ per cent. of the total output. Copper loss, 60 watts in each circuit at full load, or a total copper loss of 2 per cent. Transforming ratio, 2,000 volts to 100 at full load. Thus it will be well to make the ratio 2,000 to 102 at no load to reduce the effect of the copper drop and magnetic leakage. The frequency we will assume to be 100 complete cycles per second. The designs will be taken at random, and will not necessarily be the most economical of the various types. The cost of material may be taken as giving some indication of the probable cost of manufacture. So, assuming the copper to cost 10d. per pound, and the iron 4d., we shall work out in each case the cost and weight per kilowatt output.

First Design.—This shall be of the long shell type having a square core. This form of transformer is largely used both on the Continent and in England, but the different makers use various methods of building up the core and of winding the copper circuits. We will assume that a 4in. core is required, and that the general outline of the iron circuit is that shown in Fig. 6. The gross area of the section of the core is then 15·4 square inches, and allowing for 85 per cent. of the space being taken up by the iron, this gives a core area of 13·1 square inches, or 84·5 square centimetres. Adding up the lengths of the various parts we get a total length of 48in., which gives a volume of 630 cubic inches of iron, weighing 176lbs.

Now the total iron loss was to be 150 watts, so that the loss in hysteresis and Foucault currents for 1lb of iron is $\frac{150}{176}$ = ·85 watt. From the curve, Fig. 4, it will be seen that this corresponds to an induction of about 3,600 lines per square centimetre.

Hence the total flux $F = 3,600 \times 84\cdot5$.
$$= 304,000.$$

From this we can now by aid of (1) determine the number of turns required in both windings. For the secondary we have—

$$e_2 = 4\cdot45 \; F \; \tau_2 \; n \; 10^{-8}; \text{ and in this case}$$
$$e_2 = 102, \; F = 304,000, \; n = 100;$$

$$\therefore \tau_2 = \frac{102}{4\cdot45 \times 100 \times 304,000} \times 10^8 = 75\cdot5.$$

An even number of turns must be used for convenience, and hence 76 will be nearest to the number required. It is always well to calculate the secondary first, and to obtain an even number of turns in it, as it is not well to introduce half and quarter turns in either of the windings. We get the number of turns on the primary from the ratio of the voltages.

Thus
$$\frac{\tau_1}{\tau_2} = \frac{e_1}{e_2} = \frac{2,000}{102},$$

$$\therefore \tau_1 = \frac{76 \times 2,000}{102} = 1,490.$$

The next step to be considered is the arrangement of the copper circuit. It is found that if the coils are wound separately and placed one at either end of the core, as shown in Fig. 1, a large magnetic leakage takes place at full load. This causes a drop in the voltage, which is very objectionable, as explained above. There are several methods of obviating this defect, and one is to split the two

Alternate Current Transformers. 31

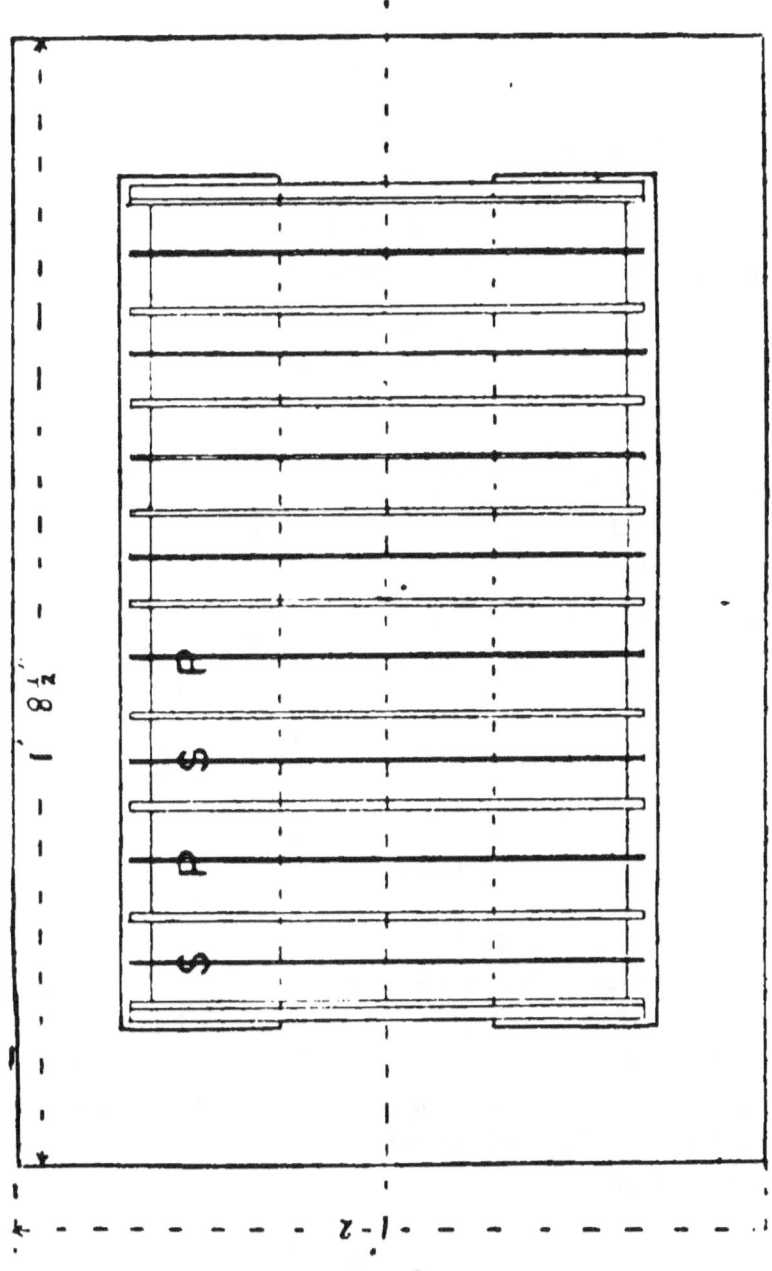

FIG. 8.

windings into sections, which are placed alternately along the core, Fig. 8. In this way the demagnetising effect of the secondary is not concentrated at one part of the core, and by the subdivision the magnetic leakage is reduced to reasonable limits. So in this transformer we will make four segments in the secondary, and interleave them with four of

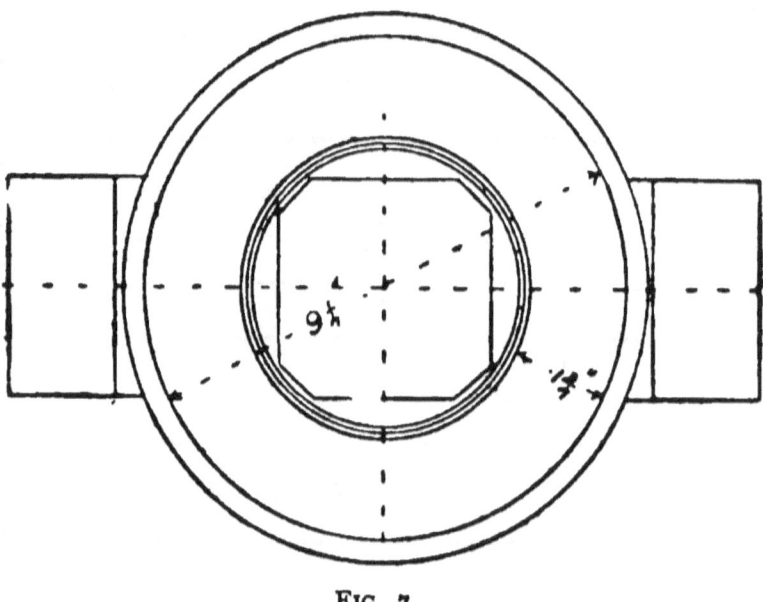

FIG. 7.

primary wire. The necessary radial insulation will be about that shown in Fig. 7, and the depth of winding we may assume to be 1¾in. The mean circumference of the winding will then be 1·97ft. = π.

Designing the primary first and using round wire for it, we get from (4)

$$d = \sqrt{\frac{i\,\tau_1\,\pi\,11\cdot 75}{\theta}}\,10^{-8}.$$

Alternate Current Transformers. 33

Here $i = 3$ amperes;
$\tau_1 = 1,490$ turns;
$\pi = 1\cdot 97$ ft. $\quad \theta = 20$ volts $= 1$ per cent.;
$l = 2,940$ ft.;

$$\therefore d = \sqrt{\frac{3 \times 1,570 \times 1\cdot 97 \times 11\cdot 75}{20}} \; 10^{-3};$$

$= \cdot 072$.

Thus, wire 72 mils diameter will give the 1 per cent. loss at full load. The insulation and clearance required in winding would bring the pitch of the wire up to about 90 mils. Dividing the depth of winding by this, we get $\frac{1\cdot 75}{\cdot 09} = 19\cdot 4$ as the number of layers which could be got in. In each segment of winding it is well to have a sub-division, so that each half can be wound separately from the bottom. This gives greater insulation, and less difference of potential between adjacent wires, than winding all in one division. So there should be $\frac{1,490}{4}$ turns to each segment of primary wire $= 372\cdot 5$, and half this number in each compartment. As it is well to work to even numbers where possible, we could wind five compartments with 19 layers of 10 turns and three compartments with 18 layers of 10 turns which would give the total number required. The actual length of core taken up by the primary wire will be $8 \times 10 \times \cdot 090 = 7\cdot 2$ in. As the mean length of turn may be altered in the final adjustments, the other details as to weight of copper, resistance etc., will be left till the actual dimensions are fixed.

Proceeding with the secondary copper, which on account of its large size should be of rectangular section we use the formulæ in group 3.

The section, S, in square inches

$$= \frac{i_2\, T_2\, \pi\, 9\cdot 2}{\theta} 10^{-6}.$$

In this case $i_2 = 60$ amperes;
$T_2 = 76$;
$\theta = 1$, and $\pi = 1\cdot 97$ as before;
so that $l = 150$ft.

$$\therefore S = \frac{60 \times 76 \times 1\cdot 97 \times 9\cdot 2}{1} \times 10^{-6} = \cdot 083 \text{ sq. in.}$$

The number of turns 76 cannot be divided exactly by 8—the number of compartments—so that we must have a different number of turns in some of them. Thus, if we wind each segment with 19 turns—one compartment having 10 turns and one nine—we shall get the total number correctly. In this case it will be best to use tape and to allow for 10 layers. Thus, $\frac{1\cdot 75\text{in.}}{10}$ give the thickness of the tape and the insulation $= \cdot 175$in. Allowing 35 mils for insulation, this leaves $\cdot 140$ in. for depth of copper and tape; ($\cdot 140$in. $\times \cdot 600$in.) will give the section required. This insulated tape will thus measure ($\cdot 175$in. $\times \cdot 635$in.), and the length of core taken up by the secondary will be $8 \times \cdot 635 = 5\cdot 1$in.

Insulating flanges are required between the segments and also between the two compartments of each segment. The space taken up by these, with the necessary winding clearance, will be about 2in.

Alternate Current Transformers.

The end flanges will be thicker, and clearance will be required between them and the yoke which will account for another inch. Thus we get for the length of core;

Primary wire............................	7·2
Secondary wire.........................	5·1
Flanges and clearances.............	3·0
	15·3

Say, 15·5in. in between yokes. This only makes the transformer half an inch shorter than the rough sketch, Fig. 6.

From Fig. 7 we see that the space allowed between core and yokes is just that required, so that the length is the only dimension needing alteration. The copper circuit calculations will therefore be correct, and will not need alteration, as might have been the case, if for instance, the depth of winding had to be increased owing to the wire taking up too great a length on the core. We will therefore proceed to carefully calculate the iron losses and magnetising current. Assuming, as before, that the total section of the two yokes together is equal to that of the core, the volume of iron = 13·1 × 47in = 616 cubic inches, or 172·5lb.

For $\tau_2 = 76$, the exact F from (1) is 302,000 C.G.S. lines, therefore $\mathfrak{B} = \dfrac{302,000}{84·5} = 3,580.$

From curve 4 this corresponds to a loss of ·84 watt per pound, so that the total iron loss is 172·5 × ·84 = 144·8 watts. From this we get that the current

required by the iron losses at open circuit $= \dfrac{144\cdot 8}{2{,}000}$
$= \cdot 0724$ ampere $= i_H$ Fig. 3.

The next step is to calculate $i\mu$, the current required by the permeability of the iron. This is got by means of the formula (2):

$$i\mu = \dfrac{\mathcal{B}\, l}{\mu\, \tau_1\, 1\cdot 76}.$$

For this transformer, l, the mean length of the magnetic path in centimetres, is 123;
$\mathcal{B} = 3{,}580$;
μ at this induction, from Fig. 5 $= 1{,}960$;
$\tau_1 = 1{,}490$;

$$\therefore i\mu = \dfrac{3{,}580 \times 123}{1{,}960 \times 1{,}490 \times 1\cdot 76} = \cdot 0858 \text{ ampere.}$$

As from Fig. 3, i_H and $i\mu$ are the two sides of a right-angle triangle, we obtain the magnetising or no-load current by taking the square root of the sum of the squares of these two. Thus:

$$\begin{aligned} i &= \sqrt{i_H{}^2 + i\mu^2}: \\ &= \sqrt{\cdot 0724^2 + \cdot 0858^2}; \\ &= \cdot 112 \text{ ampere;} \end{aligned}$$

which is $3\cdot 74$ per cent. of the full-load current in the primary.

Now completing the details for the primary copper by aid of (4), we get the resistance from

$$\begin{aligned} r &= \dfrac{\pi\, \tau_1\, 11\cdot 75}{d^2}\, 10^{-6}; \\ &= \dfrac{1\cdot 97 \times 1{,}490 \times 11\cdot 75}{\cdot 072 \times \cdot 072}\, 10^{-6}; \\ &= 6\cdot 69, \text{ say } 6\cdot 7 \text{ ohms.} \end{aligned}$$

It will be seen that when the full primary current of three amperes is flowing, this gives a loss of 20 volts as intended.

The weight is now obtained from

$$g = d^2 \pi \tau_1 \, 3 \cdot 02;$$

placing the respective values of d, π and τ_1 we get $g = 46 \text{lb}$.

The secondary is next treated in the same way by aid of (3), which applies for strip or any wire when the section is known.

Hence, $$r = \frac{\tau_2 \pi \, 9 \cdot 2}{S} \, 10^{-6};$$

or, placing in the value as determined above,

$$r = \frac{76 \times 1 \cdot 96 \times 9 \cdot 2}{\cdot 083} \, 10^{-6};$$
$$= \cdot 0166 \text{ ohm.}$$

This when multiplied by 60 amperes, the maximum current, gives us a drop of one volt. These calculations of the resistances of the primary and secondary winding check the accuracy of the previous work, in which a 1% drop was assumed.

The weight, g, equals $S \, \tau_2 \, \pi \times 3 \cdot 85$, and so

$$g = \cdot 083 \times 76 \times 1 \cdot 96 \times 3 \cdot 85,$$
$$= 47 \cdot 5 \text{lb.}$$

Now collecting the details of the design together, we get the following list:

Six-Kilowatt Transformer, Shell Type:—Ratio of transformation, 2,000/100 volts; $\frown\!\smile$ = 100.

Core, 4in. square; area of iron, 84·5 square centimetres.

$\mathfrak{B} = 3{,}580$, $F = 302{,}000$, $l = 123$.
Primary: $\tau_1 = 1{,}490$. $\pi = 1\cdot96$.

Conductor, 72-mil wire wound in four segments of two divisions as follows.

    5 divisions of 19 layers of 10 turns.
    3 ,, ,, 18 ,, ,, 10 turns.

Maximum depth of winding, $1\cdot7$in.
$r = 6\cdot7$ ohms; loss = 1 per cent.; weight, 46lb.

Secondary: $\tau_2 = 76$. $\pi = 1\cdot96$.

Conductor, strip ($\cdot140$in. × $\cdot600$in.), wound in four segments of two divisions, as follows:

    4 divisions of 10 layers of 1 turn, and
    4 ,, ,, 9 ,, ,, 1 turn.

Maximum depth of winding, $1\cdot75$in.
$r = \cdot0166$ ohm; loss = 1 per cent.; weight, $47\cdot5$lb.

Losses—Iron .... $144\cdot8$ watts, or $2\cdot42$ per cent.
    Copper, primary   60   ,,   1   ,,
    Copper, secondary 60   ,,   1   ,,

                $264\cdot8$ watts.

Therefore the efficiency at full load $= \dfrac{6{,}000}{6{,}264\cdot8}$.

                $= 95\cdot6$ per cent.

The watts wasted in the iron at no load are $144\cdot8$, and the apparent watts at no load = 2,000 volts × $\cdot112$ ampere = 224 watts.

So the power factor in this case $= \dfrac{144\cdot8}{224} = \cdot648$.

The weight of iron = $172\cdot5$lb., at 4d. per lb. = $57\cdot5$s.
   ,,   copper = $93\cdot5$lb., at 10d. per lb. = $78\cdot0$s.

        Total ...... 266lb. $135\cdot5$s

Alternate Current Transformers. 39

Dividing by the output, we find that the results have been obtained by using 44·3lb. of copper and iron per kilowatt, and at a cost of 22·6s. for the materials; also that the ratio of weight of copper to weight of iron = ·55.

The other point to be considered in the transformer just designed is the cooling surface per watt wasted at no load and also at full load. This should give an approximate idea of the temperature which the transformer will attain when left on the circuit. The investigation is, however, by no means so simple as might be expected at first sight. Suppose, for instance, we calculate the total cooling surface. The surface of the copper coils and insulation may be taken as that of a cylinder $9\frac{1}{4}$in. diameter and 15in. long, and thus $= \pi \times 9\frac{1}{4}$in. $\times$ 15in. $=$ 438 square inches. The end flanges may help to cool the coils, and hence, the surface of these should be added = 36 square inches. The iron in the yokes will add considerably to the above, the exposed surface of the iron is about 780 square inches. so that the total cooling surface will be 1,254 square inches. The watts wasted at no load are 144·8, and at full load the total loss rises to 264·8 watts. Working with these figures and the cooling surface given above, the loss comes out at

·115 watt per square inch at no load, and
·211 ,, ,, ,, ,, at full load.

These figures appear ample, if viewed in the light of dynamo magnet practice, but it must be remembered that dynamos have moving parts which create currents of air round the field magnets, tending to make the cooling surface more effective, whereas the transformer not only is motionless, but is boxed up in an iron case which prevents a free circulation of air. Thus the heat has to be transferred from the transformer to the air in the case, again from this air to the case, and finally from the case to the air outside. This is not absolutely correct, as the transformer loses heat by conduction where the yokes rest on the iron box, but still, roughly, we may expect at least two and a half times the rise of temperature that would be found on a field magnet having the same loss per square inch cooling surface.

Much of course, depends on the shape and mechanical construction of the transformer, and individual constants should be obtained from practice for the heating formula for each type.

Now, in the above calculation, we have worked with the total cooling surface and the total loss. Even if the loss were uniform through the mass this would give widely different internal temperatures according to the depth of material. It is, however, this internal temperature that causes faults in transformers by heating the insulation on the wires up to a point at which it either chars or becomes rotten. In the transformer designed above the most dangerous point would be the bottom of the coil wound in the centre of the transformer. The heat generated in the iron immediately under this would pass, to a considerable extent, up through the copper, and this heat has to be dissipated by the surface of the copper as well as the copper watts. Considering a vertical section 1in. long, the iron loss in the part of the core thus cut off would be 3·1 watts, and the copper loss in the wire would be about seven watts, giving a total loss in the section of 10·1 watts. Now, the external cooling surface is only $\pi \times 9\frac{1}{4}$in. = 29 square inches, so that the watts per square inch cooling surface in this central section rises to 0·35 at full load, whereas the mean obtained above was only 0·211. This is under the assumption that the whole of the heat generated in the section is transmitted radially to the surface, which is not absolutely correct; but if the coil is in the middle of the winding it is not likely to give off much heat to the adjacent coils which will be at much the same temperature.

In comparing the figures relating to the cooling of transformers the hottest point of the copper circuits should always be considered, and not the mean external temperature only. The highest temperature in the iron does not affect us much, as no ill effects occur from this cause unless the hottest part of the iron is close to the copper, which is seldom the case. The yokes in this design have the largest amount of surface per watt wasted, and hence keep cooler than the centre core. Some makers have reduced the section of the yokes for this reason, as weight can be saved by so doing and the temperature still kept normal. This, of course, causes a higher induction in the yoke, and as the curve of iron loss rises rapidly with the induction, the watts per pound will increase more rapidly than the weight diminishes. Hence the total iron loss is increased by making the yokes of smaller section. Again, working in the inverse direction, some makers have put more section of iron in their yokes than in the core. This certainly lowers the iron loss slightly, but, as will be found by actual trial, the effect is not

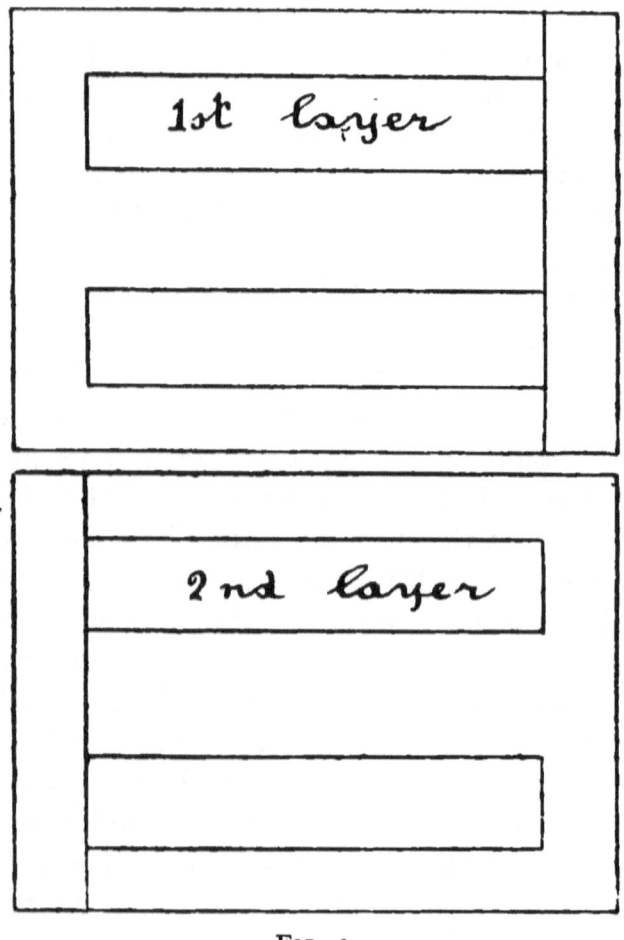

FIG. 9.

very marked, and, perhaps, is hardly worth the increase in labour and material charges.

The methods of constructing the iron cores for

Alternate Current Transformers.

transformers of proportions similar to those used in the above designs are very numerous. Almost every maker has a different way, and the advantages and disadvantages of the various methods are mainly those of cost and convenience of manufacture. In England the plates are usually arranged so as to interleave one with the other, so that each successive plate added, covers up the joint in the preceding layer. There are a large number of different ways of doing this with the present type, Fig. 9, shows one method and the reader will have no difficulty in devising many others. In this way the magnetic resistance is practically reduced to that offered by the iron, and the air resistance is negligible.* So in the curve of permeability, Fig. 5, the necessary data were obtained from a transformer having its core built up on the system, and the effect of small air resistance is thrown into the magnetic resistance of the iron. In some cases, however, the different parts of the iron circuits are fitted together at faced joints, and then the air resistance of those joints will make a considerable addition to the current marked $i\mu$, Fig. 3. The same formula as used for determining the $i\mu$, required by the iron (2) can be used for finding out the increase of current due to the air-gap, which may be called

---

* This will be readily seen when looking at the case worked out below for an air-gap having an induction of 3,000 in it. By the interleaving method this induction can easily be reduced to one line per square c.m., which would reduce the current required for the gap to a figure negligible as compared with that required by the iron, $i.e.$, ·000,191 ampere as compared with ·0,858 ampere for the iron.

In addition to this the total gap is much less in the interleaved type and will again reduce the effect.

$i\mu_a$—viz., $i\mu_a = \dfrac{\mathfrak{B}}{\mu\, \tau_2\, 1\cdot 76}$.

In this case $\mathfrak{B}$ = the induction in the air-gap.

l = the length of the air-gap in c.m.

the μ for air = 1.

The transformers made at the Oerlikon Works, in Switzerland, have the yokes fitted on to the core by faced joints, as shown in Fig. 10. Applying this plan of building up the iron circuits to the design

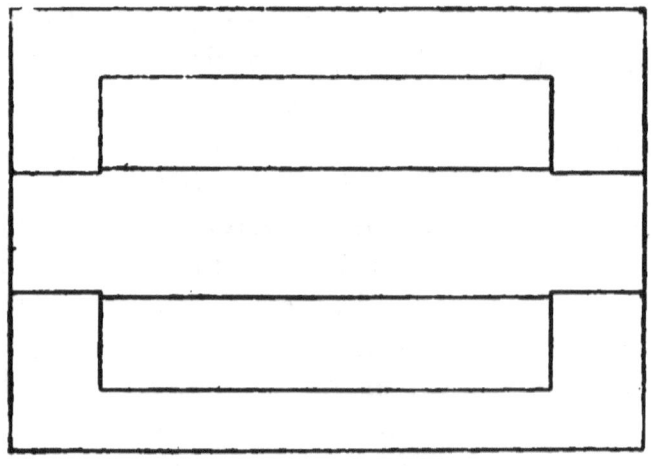

FIG. 10.

above, we get two air-gaps in the magnetic path whose combined length may average, say, ·05 c.m. The section at the joint will be larger than that of the iron in the core, say, in the present case, it reduces the induction to 3,000.

Then $i\,\mu_a = \dfrac{3{,}000 \times ·05}{1 \times 1{,}490 \times 1·76} = ·0574$ ampere.

The μ current due to the iron alone as obtained above = ·0858 ampere. The total $i\mu$ therefore = ·1432 Working out the total no-load current as before,

$$i = \sqrt{·0724^2 + ·1432^2},$$
$$= ·161 \text{ ampere, or } 5·37 \text{ per cent.}$$

So we see that the effect of the joints is to raise the

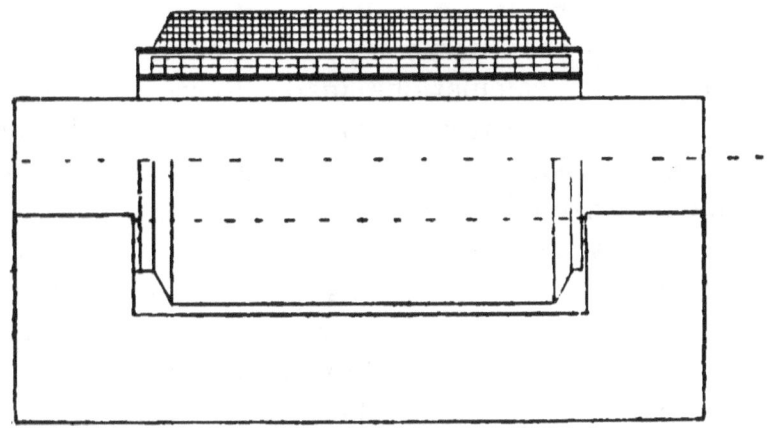

Fig. 11.

no-load current from 3·74 to 5·37 per cent. of the total full-load current. The average length of air-gap as taken above may be rather excessive, but it is extremely difficult to get good plane surfaces on the cores and yokes which are built up of thin laminations. Hence, although the faced joints reduce the labour charges on the transformer, they have almost been abandoned in England on account of the increase caused by them in the no-load current.

Fig. 11 shows this type of transformer as made by Messrs. Brown and Boveri, of Baden. As will be seen the two yokes are in this case merged into one, and the system of winding is different to that worked out in the design above. The primary and secondary wires are wound on insulating cylinders separately, and slipped on individually over the core. This method is also effective in preventing the magnetic leakage, and it enables the labour on the transformer to be sub-divided. This sketch simply shows the Brown type of mechanical detail applied to the calculations made above, and does not give the relative proportions adopted by Mr. Brown. The clamping arrangements are left out for the sake of clearness.

The next steps to be taken in order to obtain the best design of this type are as follows: (a) Design transformers with slightly smaller and larger cores and compare the figures obtained with those above; (b) next try the effect on the best design of shortening the core and winding with a greater depth of copper; if this step improves the design without materially decreasing the cooling effect, it should be carried further. The practice thus obtained in quickly running out the leading details of the various parts will soon enable the designer to gain a general idea of likely methods of improvement, and also give him an insight into the capabilities of the type. These calculations would, however, take too much space, so we will proceed to design another type of transformer in which the relative shapes of the copper and iron circuits are the exact opposites to those above.

CHAPTER II.

Second Design.—The transformer just designed had a comparatively long iron circuit of small section, with the copper wound round it. Hence the l for the iron was large, and the mean perimeter for the copper circuit small.

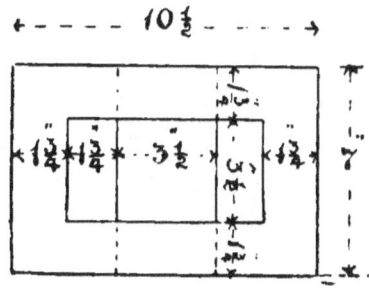

FIG. 12.

The next design shall be of the type associated with Mr. W. M. Mordey, in which the path of the magnetic lines is short, but the copper circuit has a comparatively large perimeter. We will assume, as before, a definite size of core, Fig. 12, and then afterwards alter the shape if necessary. The relative proportions of the core plates

FIG. 13.

and yokes are determined by mechanical considerations. The stampings consist of hollow rectangles of such dimensions that the pieces removed from the centre will bridge the two long sides, and thus form the centre core, Figs, 12, and 13. It will be seen that with this method of building up the iron circuit there will be space wasted between the adjacent plates, except just at the end of the core, and only half the quantity of iron can be got in that would be if the plates were arranged to break joint as in Fig, 9. Hence the Mordey transformer, although made in the smaller sizes of the general proportions shown, is now built up on a different principle.

Making the transformer 3ft. long, and allowing for 85 per cent. of the space being taken up by the plates, we get for the volume of iron :—

Alternate Current Transformers.

$\{(7 \times 10\frac{1}{2}) - (3\frac{1}{2})^2\}$ 36 × ·85 = 1,880 cubic inches, which will weigh 526lb.

The loss we assumed in the iron is 150 watts, which gives $\frac{150}{526}$ = ·285 watt per pound. From the curve, Fig, 4 this corresponds to a flux of 1,700 lines per square centimetres. Now the area of the centre core of this transformer is $3\frac{1}{2}$in. × 36in. × ·85 = 107 square inches = 690 square centimetres, so that the total flux will be

1,700 × 690 = 1,172,000 C.G.S. lines.

We can now proceed to design the secondary winding. Using formula (1)

$$e_2 = 4\cdot 45 \ F \ \tau_2 \ n \ 10^{-8},$$

we get

$$102 = 4\cdot 45 \times 1,172,000 \times \tau_2 \times 100 \times 10^{-8}.$$

$$\therefore \tau_2 = \frac{102}{4\cdot 45 \times 1\cdot 172}$$
$$= 19\cdot 6.$$

We must have an exact number of turns, as with this make of transformer the wire can only be got at at the ends of the core, and may assume 20 turns as the nearest number to the above. In this design there is as much likelihood of getting magnetic leakage as in the last, but for the present it will be sufficient to calculate the winding for two coils only, wound side by side—*i.e.*, one primary and one secondary; and then finally we must leave sufficient clearance to allow for further sub-division if necessary.

The mean perimeter of both windings will therefore

be the same, and will be about 7·5ft. The number of turns of the primary is obtained from the ratio of voltage $\dfrac{T_1}{T_2} = \dfrac{e_1}{e_2}$

$$\therefore T_1 = \dfrac{2{,}000 \times 20}{102} = 392.$$

The primary wire required to give the 1 per cent. drop at full load we get as before from (2)

$$d = \sqrt{\dfrac{i_1\, T_1\, \pi\, 11{\cdot}75}{\theta}}\; 10^{-3},$$

and here

$$i_1 = 3 \qquad \pi = 7{\cdot}5$$
$$T_1 = 392 \qquad \theta = 20$$
$$\therefore d = \sqrt{\dfrac{3 \times 392 \times 7{\cdot}5 \times 11{\cdot}75}{20}}\; 10^{-3}$$
$$= {\cdot}0721.$$

Say 72-mil wire, covered with insulation to a diameter of 87 mils.

In this case the space for winding is absolutely fixed by the economic proportions of the iron circuit. So it is necessary that the wire shall be such that it can be wound easily in the spaces left.

Assuming a minimum insulation of ⅛in. everywhere between copper and iron, the depth of winding may be 1·5in. and $\dfrac{1{\cdot}5}{{\cdot}087} = 17{\cdot}3$, so that we may use 17 layers of the above wire.

To get the 392 turns we then require 23 turns per layer, so the length of the coil when wound will be

$$23 \times {\cdot}087 = 2\text{in}.$$

Alternate Current Transformers.

The secondary will be best wound as tape in 10 layers of two turns each. The depth of the insulated tape must be one-tenth of the total depth available. $\frac{1\cdot5\text{in.}}{10} = \cdot150\text{in.}$, and allowing 40 mils for insulation, the thickness of the tape will be ·110in. The necessary section is obtained by (3) from

$$S = \frac{i_2 T_2 \pi \times 9\cdot2}{\theta} 10^{-6},$$

and here

$$i_2 = 60 \qquad \pi = 7\cdot5$$
$$T_2 = 20 \qquad \theta = 1.$$
$$\therefore S = \frac{60 \times 20 \times 7\cdot5 \times 9\cdot2}{1} 10^{-6}$$
$$= \cdot083 \text{ square inch.}$$

Therefore the tape must be $\frac{\cdot083}{\cdot110} = \cdot75\text{in.}$ broad, and will be of the following size: Bare, ·750in. × ·110in.; insulated, ·790in. × ·150in. The length required for two turns placed side by side will be ·790in. × 2 = 1·580in., say, 1·6in. Now the primary winding took 2in., so that for wire alone the length of window required would be 2in. + 1·6in. = 3·6in., whereas 3·5in. is the total length available for insulation and wire—see drawing, Fig. 12.

Hence with the core as sketched there is not room for windings giving a 1 per cent. copper loss in both primary and secondary.

There are two courses open if the iron parts are not to be altered. The one is to increase the copper loss by using thinner wire, which can be wound in the space, and the other is to decrease the number of turns in each winding. This will increase the induction, and hence the iron loss. We can then, however, leave the total copper loss at two per cent. On the whole, perhaps, the latter course is preferable, as the regulation of the transformer is impaired by allowing more copper loss. The smaller number of turns gives a correspondingly shorter length of copper in each circuit, and hence conductors of smaller sectional area can be used, which again decreases the space required. If we fix τ_2 at 18 turns we shall be likely to get the copper in the space available—3·5in. × 1·75in. The number of turns in the primary, τ_1, will therefore be

$$\tau_1 = \frac{2{,}000 \times 18}{102} = 354.$$

The whole of the copper circuits must be redesigned to these figures.

Taking the primary first and replacing the old number of turns by the figure 354, we get

$$d = \sqrt{\frac{3 \times 354 \times 7\cdot5 \times 11\cdot75}{2}}\, 10^{-3}$$
$$= \cdot0684.$$

Wire 68 mils diameter may therefore be used, which will have a diameter when insulated of about 83 mils. The number of layers will be $\frac{1\cdot5\text{in.}}{\cdot083\text{in.}} = 18$, and 20 turns per layer would give us 360 turns. So the last layer must be wound with only 14 turns in order that the total number may be 354.

Alternate Current Transformers.   53

The length of the winding space required for the primary will be 20 × ·083 = 1·66in., say 1·75in., should be allowed. The weight of primary will be got as before from

$$g = d^3 \pi \tau_1 \, 3\cdot 02.$$
$$= \cdot 068^2 \times 7\cdot 5 \times 354 \times 3\cdot 02.$$
$$= 37 \text{lb}.$$

From the formula for the resistance we get $r = 6\cdot 73$ ohms, which, with three amperes, gives slightly over the 20-volt drop allowed at full load.

Now, in the secondary, there are only 18 turns, which can be wound in nine layers of two turns each. The section in square inches of the conductor from

$$S = \frac{i_2 \tau_2 \pi \, 9\cdot 2}{\theta} 10^{-6};$$

becomes equal to $\dfrac{60 \times 18 \times 7\cdot 5 \times 9\cdot 2}{1} 10^{-6};$

$$= \cdot 0745 \text{ square inch}.$$

The dimensions of the tape must be determined as above to suit the space available. The insulated tape will have a thickness of $\dfrac{1\cdot 5}{9} = \cdot 167$in. The tape uninsulated may be made ·125in. thick, and then a breadth of ·6in. gives the required section. So tape (·600in. × ·125in.) will be used covered to (·640in. × ·165in.); the two turns side by side requiring a space of 1·28in., say 1·3in. So in this design we need a total length in the window of 1·3in. + 1·66in. = say 3in. for wires, leaving ½in. for insulation. This will be ample, and hence these conductors can be finally adopted. The weight of the secondary will be from (3)—

$$g = S\,\tau_2\,\pi \times 3\cdot 85\,;$$
$$= \cdot 0745 \times 18 \times 7\cdot 5 \times 3\cdot 85\,;$$
$$= 38\cdot 7 \text{lb}.$$

and the resistance worked out from

$$r = \frac{\tau_2\,\pi \times 9\cdot 2}{S}\,10^{-6}\,;$$

$$= \frac{18 \times 7\cdot 5 \times 9\cdot 2}{\cdot 0745} \times 10^{-6}\,;$$

$$= \cdot 0167 \text{ ohm}.$$

FIG. 14.

The arrangement of the winding as designed is shown in Fig. 14, but we may find after further consideration that it is not the best method of winding when the question of magnetic leakage is considered.

The next step is the accurate determination of the iron loss and the no-load current. The induction

Alternate Current Transformers.

assumed at first will be altered, owing to the change in the number of turns. From (1) we get that

$$e_1 = 4{\cdot}45\, F\, \tau_1\, n\, 10^{-8},$$

and now e_1 = 2,000, and τ_1 = 354.

$$\therefore 2{,}000 = 4{\cdot}45 \times F \times 354 \times 100 \times 10^{-8};$$

$$F = \frac{2{,}000}{4{\cdot}45 \times 354}\, 10^{-6}.$$

$$= 1{,}270{,}000.$$

The area of the core being 690 square centimetres, $\mathfrak{B}$ = 1,840, instead of 1,700 as first assumed. At this induction the loss per pound of iron is ·32 watt.

The total weight of iron being 526lb., we get a loss of 526 × ·32 = 168 watts, or 2·81 per cent. For this loss, at 2,000 volts, a current of ·084 ampere will be required, which we call i_H, Fig. 3.

The data for determining the current required to magnetise the iron is got from formula (2)—

$$i_\mu = \frac{\mathfrak{B}\, l}{\mu\, \tau_1\, 1{\cdot}76}.$$

Here $\mathfrak{B}$ = 1,840; l, the mean length of magnetic path, is 17in. = 43·2 centimetres. The μ corresponding to the above induction for this iron is (from Fig. 5) 1,280, and τ_1 = 354.

$$\therefore i_\mu = \frac{1{,}840 \times 43{\cdot}2}{1{,}280 \times 354 \times 1{\cdot}76}$$

$$= 0{\cdot}0995.$$

The no-load current = $\sqrt{i_H^2 + i_\mu^2}$;

= $\sqrt{({\cdot}084)^2 + {\cdot}0995^2}$;

= ·130 ampere;

= 4·34 per cent. of the full-load primary current.

The various details of the design are therefore as follows:

Six-Kilowatt Transformer—Second Design.—Ratio of transformation, 2,000/100 volts; $\frac{\frown}{\smile}$ = 100.

Core, 3½in. × 36in.; area of cross-section of iron, 690 square centimetres.

$\mathfrak{B}$ = 1,840; F = 1,270,000; l = 43·2cm.

Primary: τ_1 = 354; π = 7·5ft.

Conductor, 68-mil wire insulated to 83 mils, and wound in 18 layers of 20 turns.

r = 6·73 ohms, hot; loss = 1 per cent.; weight, 37lb.

Secondary; τ_2 = 18; π = 7·5.

Conductor, tape (·600in. × ·125in.) insulated to (·640 × ·165) wound in nine layers of two turns each.

r = ·0167 ohm; loss, 1 per cent.; weight 38·7lb.

Losses—Iron............	168 watts,	or 2·81 per cent.
Copper, primary...	60 ,,	or 1·0	,,
Copper, secondary	60 ,,	or 1·0	,,
Total	288 watts	4·81 per cent.

Therefore the efficiency at full load is

$$\frac{6,000}{6,288} = 95\cdot4 \text{ per cent.}$$

The watts wasted at no load are 168, and the apparent watts at no load are 2,000 × ·130 = 260.

Therefore the power factor will be $\frac{168}{260}$ = ·646.

The weight of iron = 526 lb., at 4d. per lb. = 175·5s.
,,   copper = 75·7lb., at 10d. per lb. = 63·1s.

Total ...... 601·7lb.   238·6s.
= 100lb. and 39·8s. per kilowatt.

Alternate Current Transformers. 57

The ratio of the weight of copper to weight of iron is ·144.

The iron in this transformer provides the greater part of the cooling surface, and the laminations as placed act well as radiators. The total surface of iron exposed to the air is 1,358 square inches, and the ends of the copper coils, with insulation, provide about 132 square inches more. This gives a total cooling surface of 1,490 square inches. The watts wasted at no load are 168, and at full load the total loss is 288 watts. These figures give a mean of ·113 watt per square inch at no load, and ·193 watt per sqare inch at full load.

As might have been expected, from the large amount of material used, these figures are better than those obtained in the first design in spite of the increase in the iron loss. Besides this, the plates surrounding the coils are arranged so that the laminations all run out to the surface. Thus the heat can be readily dispersed by conduction. Still, in spite of this great advantage, which accounts in some measure for the low temperature rises found in the Westinghouse and Mordey transformers, it must be remembered that it is impossible to obtain the temperature of the wire in the centre of the transformer. The heat generated in wires at the centre will mostly pass out through the insulation to the iron and thence to the air. Hence the temperature of the wire in the middle must be considerably higher than that of the inside edges of the iron plates. The inside edges will again be hotter than the outside, but the difference will not be so much as the iron is a good

conductor compared with the insulation on and round the copper. So although the temperature on the outside is low, the copper wires in the centre may get hot.

The great point to be noticed in the above results from the design is, that the ratio of weight of copper to weight of iron is small, (only ·144). This ratio is fixed by the mechanical arrangements of the core, but, as seen above, this method of construction has already been modified in order to get more iron in. The saving in the iron stamping effected by the use of every piece of iron stamped out will not make up for the want of efficiency in the transformer, and so for large sizes the proportions given above are not adhered to. By enlarging the windows we are enabled to wind more turns in both primary and secondary, and hence to reduce the flux required. Again, by this alteration less iron can be used, and hence the total weight and cost can both be reduced without prejudicing the efficiency. Still, the design worked out above would give good results if better iron were used.

Another point to be noticed, is that the power factor is of the same order as that obtained in the first design, where the length of the iron circuit was 123 centimetres. The decrease in length of the path of the magnetic lines would give a corresponding decrease in the current required to magnetise the iron, $i\mu$, if the number of turns in the primary were the same. In this case, however, the number of turns in the primary are much fewer than in the previous design, and hence the $i\mu$ comes out about the same proportion of the total no-load current. The value of i_H is independent of the shape of the iron circuit, and depends only on the induction, frequency, and the total weight of iron used, provided the plates are so arranged that no secondary currents are induced in them other than the usual Foucault currents. In further development of this design, the two alterations most likely to effect improvements in the results are (a) the enlargement of the window, (b) the shortening of the whole transformer. How far the reduction of the length can be carried without increasing the breadth of core is a matter to be determined by careful trial.

The mechanical construction of the above transformer with the original method of building the core is as follows. The two coils are wound on a former of the exact shape required by the windows left in the iron, and are then carefully insulated. The ends of the wires are usually all brought out at same end of the transformer. The semi-circular spaces at each end of the coils are filled in with a block of wood or non-conducting material, Fig. 13. The coils are then taken off the former. In building up the iron on the insulated coils, a hollow rectangular stamping is first threaded over them, and next the piece forming the core is placed through the coil, as shown in Fig. 12. These operations are repeated till the whole length has been built up. Bolts connected to flanges at either end are tightened occasionally, so that the plates are forced into as small a length as possible. When the full number of the stampings have been got in, these bolts are finally tightened, and thus prevent any displacement of the plates.

There are several other makers who use a type

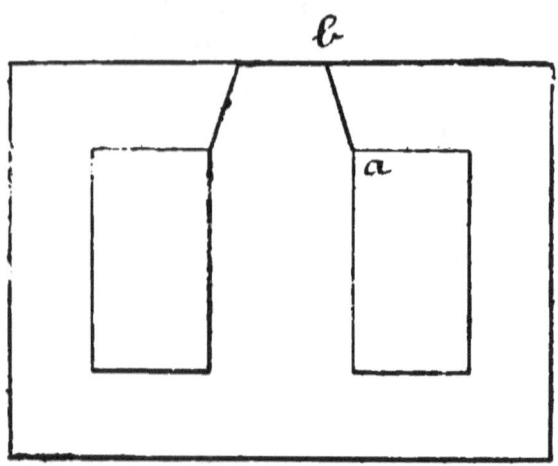

FIG. 15.

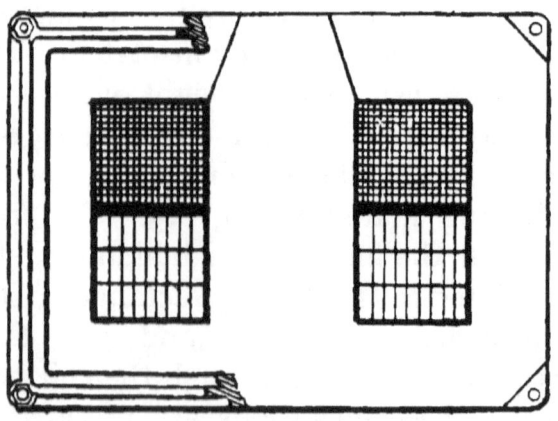

FIG. 16.

similar in general outline to the above, and the mechanical constructions adopted by two of them are worthy of special note. The Westinghouse transformer is shown in section in Figs. 15 and 16.

As will be seen, the appearance of the section is very similar to that of the transformer, Fig. 12, except that the windows take out a larger proportion of iron; also the length of this transformer, as compared to its other dimensions, is shorter. The splendid efficiency of a 6·5 kilowatt transformer of the Westinghouse type, as measured by Dr. J. Hopkinson and Dr. Fleming (see the list, page 11) makes this design of special interest. The material used in the core is said to be a special make of very soft steel plate. The core plates consist of single stampings, Fig. 15, of rectangular shape, with two windows punched out where the wire is to be placed. To enable these stampings to be threaded on to the coils, the plate is cut through in two places, from the window out to the long edge of this rectangle, as shown by the lines (*a b*), Fig. 15. The cross yokes, thus separated from the core, can now be bent back, and the stamping placed in position on the coils. These cross-pieces are then straightened again, and the stamping embraces the coils. Alternate plates are placed on from different sides, so that the joints (*a, b*) on one plate are covered by an unbroken part of the next. The rest of the construction is carried out as before, the wires being wound on formers, and well insulated from the iron by insulating cloth and other material able to withstand the mechanical abrasion while the core is being built up. When complete, the iron plates are clamped up by four bolts placed at either corner, as shown in Fig. 16. The good quality of the iron or steel plates used accounts in some measure for the splendid efficiency of this transformer, but the general proportions must also be very carefully designed.

The other transformer referred to is the Elwell-Parker, and in this case the iron circuit round each of the windows is kept distinct. The coils are insu-

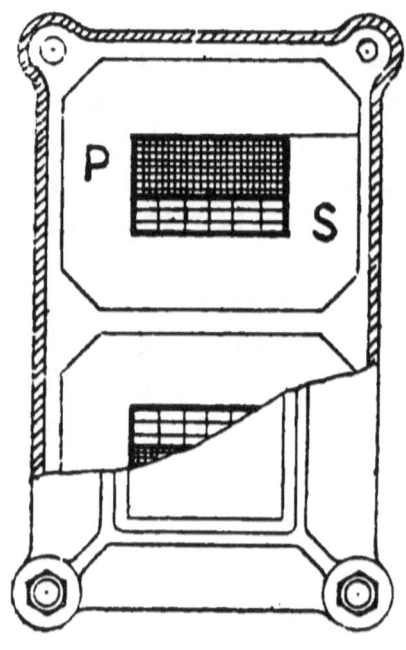

FIG. 17.

lated in the usual manner, and each side of the long rectangle of copper wires thus formed are surrounded by plates of iron, Fig. 17. These plates are hollow rectangles, stamped out whole and then cut through at one place, much in the same manner as in the Westinghouse plates. This enables one side to be bent back, in order that the stamping may be slipped

Alternate Current Transformers. 63

on to the wire. The successive plates are arranged to break joint as described in above for other designs. The transformer of this make, shown in Figs. 17 and 18, is much shorter in proportion in its length than the design worked out, and as it is to scale a rough idea of the proportion of copper to iron can be obtained.

It will be noticed that all the makers of the general

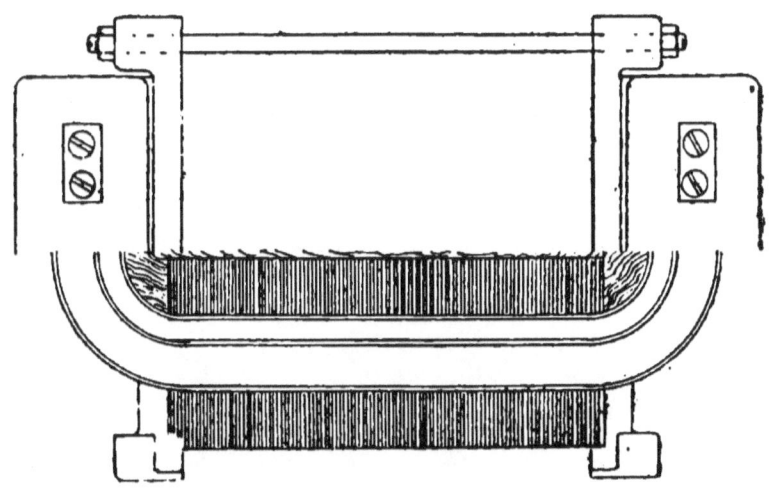

FIG. 18.

type illustrated by this second design seem inclined to keep the length of the transformer down. The object of this is that the mean length of the copper circuits to enclose a given area, or take a given total flux, is much lessened by this step. This reduction in the π of the copper saves weight and space, both of which are valuable. Still, some makers do not reduce the copper weight, but prefer instead to add more turns in the space, and thus reduce the iron loss. This seems an easy method of increasing the efficiency, but the magnetic leakage is also much increased unless further changes in the methods of arranging the copper circuits are made. That this is so is seen from the fact that the magnetic leakage in the Westinghouse transformer in the list is its worst feature. It is 1·02 per cent., and is the highest percentage leakage recorded by Dr. Fleming.

CHAPTER III.

❖❖❖❖❖❖❖❖❖❖❖❖

Third Design.—In this, the last transformer to be designed, we will make the mean lengths of the copper and iron circuits nearly equal. The first design had a short copper circuit and a long iron core; then, in the second, these relative proportions were reversed and the iron circuit was short. Now, the mean length of the turns in the copper circuits and the length of the path of the magnetic lines will be, roughly, the arithmetical means of the respective values in the two previous designs. This gives a transformer of the type used by Ferranti, in which a rectangle core of iron is surrounded by rectangular formers containing the copper conductors.

Assuming the core to be 9in. long by 3in. broad, and the length of winding formers to be about 9½in., we can roughly estimate the weight of iron in the transformer, Figs. 19 and 20. The cross-section of the iron equals 85 per cent. of 27 square inches = 23 square inches, or 148 square centimetres. The length of the iron circuit cannot be accurately

66   *The Design of*

determined till the winding has been fixed, but will be, roughly, 28in., so that the volume of the iron = 23 × 28 = 642 cubic inches and weighs 180 lbs. The iron loss allowed is 150 watts, so that the loss per

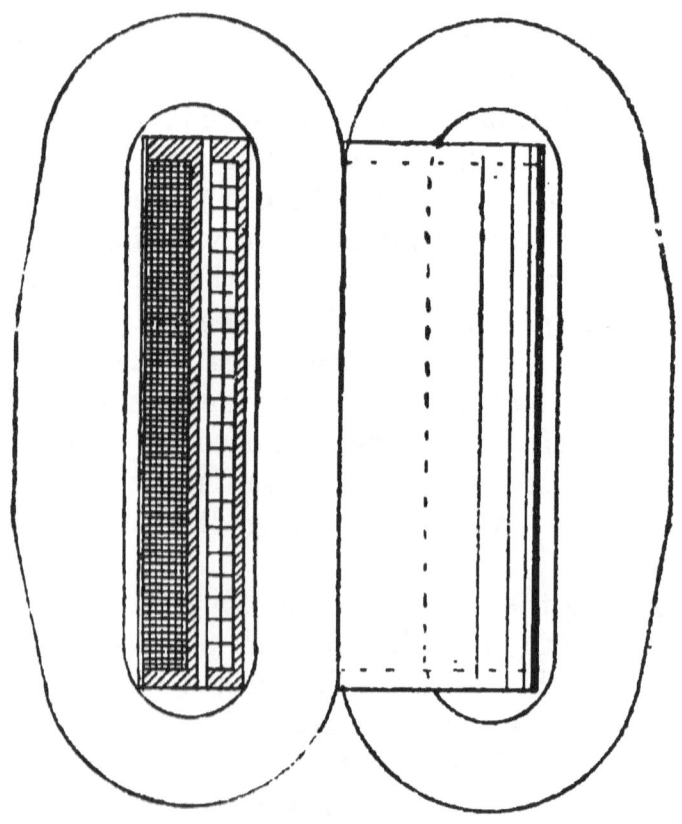

Fig. 19.

pound $= \frac{150}{180} = $ ·835watt. The curve of total iron losses, Fig. 4, shows that this corresponds to an induction of about 3,500 lines per square centimetre. The core area being 148 square centimetres, hence

the total flux equals 148 × 3,500 C.G.S. lines,
∴ F = 518,000.

Designing the secondary winding first from (1)
$$e_2 = 4\cdot 45 \; F \; \tau_2 \; \iota \; 10^{-8},$$

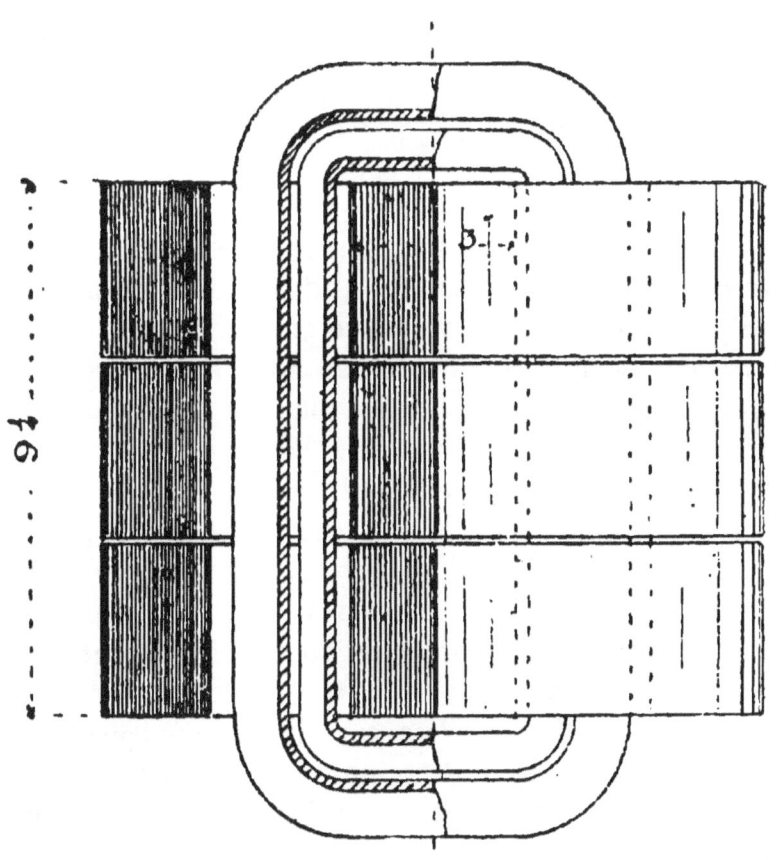

FIG. 20.

and allowing 102 volts to make up for the drop due to copper loss at full load, we get

$$102 = 4\cdot 45 \times 518{,}000 \times \tau_2 \times 100 \times 10^{-8};$$

$$\tau_2 = \frac{102}{4\cdot 45 \times \cdot 518} = 44\cdot 2 \text{ turns.}$$

We must have an even number of turns, as two layers will be the most convenient method of arranging the secondary strip—44 being the nearest even number, will do well for the number of turns in the secondary. The primary turns can now be calculated from

$$T_1 = \frac{T_2 \times 2{,}000}{102} = \frac{44 \times 2{,}000}{102},$$

$$= 865$$

In this type of transformer the coils are usually wound on separate formers and slipped on individually. The question of which should be nearer the core has been often discussed, and the balance of advantage appears to lie in placing the secondary next to the iron, as then the high volt winding is farther from the iron, and less likely to leak to earth.

Also the thick wire of the secondary is wound on a smaller former, and is hence more stable. The question of eddy-currents in the copper due to leakage of magnetic lines from the core out through the copper has been urged as a reason for putting the thick wire outside. With certain strained assumptions as to leakage and copper weight, this appears to be a valid reason but for the fact that the eddy-current losses in the copper are so small as to be negligible in most transformers. Assuming the leakage to take place radially from the core, the induction and hence the eddy current loss per unit volume of copper, is greater near the core. But the volume of copper required increases more rapidly than the perimeter of the winding if a constant full-load drop is allowed, and so the increase of volume varies almost with the square of the perimeter. This increase in volume or weight of conductor balances the decrease in the density of the leakage field, and keeps the Foucault loss constant. Hence there is no disadvantage from leakage caused by winding the thick wire near the core. Placing, therefore, the secondary inside, it is necessary to design it first, as its thickness will affect the perimeter of the primary.

The clearances between the insulating cylinder and the iron must be slightly more than in the previous design, owing to the method of core construction used. Fig 20, shows the general outline of the coils, and from it the mean perimeter of the secondary equals 2·37ft. The total length of the insulating cylinder being 9½in., we shall get about 8½in. for winding space. The section of the tape to be used is got from (3)

$$s = \frac{i\, \tau_2\, \pi_2\, 9\cdot 2}{\theta} 10^{-6};$$

and $i = 60$ $\qquad \pi_2 = 2\cdot 37$ft.
$\tau_2 = 44$ $\qquad \theta = 1.$
$\therefore s = 60 \times 44 \times 2\cdot 37 \times 9\cdot 2 \times 10^{-6};$
$= \cdot 0572$ square inch.

The tape had best be wound in two layers of 22 turns each, so that the insulated tape should measure $\frac{8\cdot 5\text{in.}}{22}$, say, ·385in., in order that 22 turns may be wound in 8·5in. Uninsulated the copper would be about ·350in. broad, and to give the section must be ·165in. thick. Tape (·350in. × ·165in.) will therefore be used, and the two layers will give a total depth of winding of ·4in.

Calculating the resistance as in previous designs for the exact section of the tape, ·0577 square inch, we get from (3) that $r_2 = ·0166$, which, with 60 amperes, gives just under the one volt drop allowed.

The weight of copper from
$$g = S \pi \times 3·85,$$
by substituting the proper value for the symbols becomes, $g = ·0577 \times 14 \times 2·73 \times 3·85;$
$$= 23 \text{lb}.$$

The clearance between the outside of this secondary winding and the primary cylinder must be sufficient to allow the latter to be slipped on. The space left is by no means wasted, as it provides ventilation to the wires in the centre of the coil. In fact, with a transformer so placed that these cylinders are vertical, the chimney action in these air spaces between the windings is most useful in cooling the wires. This space and the necessary thickness of cylinder and wire raises the mean perimeter of the primary wire to about 2·87ft. This is the only remaining unknown quantity in the equation for the diameter of the wire:

$$d = \sqrt{\frac{i_1 \tau_1 \pi_1 \, 11·75}{\theta}} \, 10^{-3},$$

the other values being as follows:
$$i = 3; \tau_1 = 865; \pi = 2·87; \theta = 20;$$
$$\therefore d = \sqrt{\frac{3 \times 865 \times 2·87 \times 11·75}{20}} \, 10^{-3},$$
$$= ·066.$$

Alternate Current Transformers.

So 66-mil wire will give the 1 per cent. loss, and it can be wound in eight complete layers of 100 turns and one layer of 65 turns. In between each layer it will be necessary to place a sheet of insulating material, as there will be a difference of potential of about 470 volts between adjacent wires in different layers at the ends of the coils. This extra insulation will increase the depth of the winding to about ·85in. The resistance of the wire from (4) is

$$r = \frac{T_1 \pi_1 11 \cdot 75}{d^2} 10^{-3},$$

$$= \frac{865 \times 2 \cdot 87 \times 11 \cdot 75}{\cdot 066 \times \cdot 066} 10^{-3},$$

$$= 6 \cdot 68 \text{ ohms.}$$

The weight of conductor, from the formula following the above, works out to 32·7lb. These calculations complete the design of the copper circuits.

With the Ferranti system of building up the core and yoke, the question of winding space does not affect the calculations so much. It is necessary first to design the coils to suit the core area allowed, and then to get out the length of plates required to encircle these coils with ample clearance. Leaving the actual details of construction till later, we can, from Fig. 19, get the mean length of the iron circuit, which is 28·5in., or 72·7 centimetres. The cross-section of iron was 23 square inches, so that the exact volume will be 656 cubic inches, weighing 183 lb. The total field, F, required when $\tau_2 = 44$ will be

520,000 C.G.S. lines, so that $\mathfrak{B} = \dfrac{520,000}{148} = 3,520$. The loss per pound in watts at this induction for the iron assumed by curve Fig. 4, is ·827, so that the total iron loss = 827 × 183 = 151 watts. The watt current, i_{II}, required to give this power at 2,000 volts = $\dfrac{151}{2,000}$ = ·0756 ampere.

The next step is to determine the μ current from $i\mu = \dfrac{\mathfrak{B}\, l}{\mu\, \tau_1\, 1·76}$.

In this transformer $\mathfrak{B}$ = 3,520;
l = 72·7 centimetres;
μ = 1,950 from curve, Fig. 5;
τ_1 = 865;
$\therefore i\mu = \dfrac{3,520 \times 72·7}{1,950 \times 865 \times 1·76}$;
= ·0864 ampere.

Now i, the magnetising or no-load current,
$= \sqrt{i_{\text{II}}^2 + i\mu^2}$,
$= \sqrt{·0756^2 + ·0864^2}$,
= ·115 ampere,

which is 3·83 per cent. of the full-load primary current.

Collecting all the figures into a list.

Six-Kilowatt Transformer—Third Design.—Ratio of transformation, 2,000/100 volts; $\frown\smile$ = 100.

Core, 9in. × 3in.; area of cross-section of iron, 148 square centimetres.

$\mathfrak{B}$ = 3,520; F = 520,000; l = 72·7cm.
Primary: τ_1 = 865; π = 2·87ft.

Wire, 66 mils diameter, wound in eight layers of 100 turns per layer and one layer of 65 turns per layer.

$r = 6·68$ ohms, hot; loss = 1 per cent.; weight, 32·7lb.

Secondary: $\tau = 44$; $\pi = 2·73$.

Conductor used, tape (·350in. × ·165in.), wound in two layers of 22 turns each.

$r = ·0166$ ohm, hot; loss, 1 per cent.; weight, 23lb.

Losses—Iron .... 151 watts ... 2·51 per cent.
    Copper, primary ... 60 ,, ... 1·0 ,,
    Copper, secondary 60 ,, ... 1·0 ,,

                      271 watts   4·51 per cent.

The efficiency at full load being

$$\frac{6,000}{6,271} = 95·7 \text{ per cent.}$$

Magnetising current = 3·83 per cent., so that the load factor, or ratio of true watts to apparent watts on open circuit, equals ·657.

The weight of iron = 183lb. at 4d. per lb. = 61s.
   ,,   copper = 55·7lb. at 10d. per lb. = 46·5s.

    Total .... 238·7lb.   107·5s.

Which gives 39·8lb. and 18s. per kilowatt output.

The ratio of weight of copper to weight of iron equals $\frac{55·7}{183} = ·304$.

The effective cooling surface of this transformer is difficult to estimate correctly, and comparisons made on the basis of the watts wasted per square inch total external surface will not afford a true idea of the heating limits. This is due to the number of ventilation spaces running right through both the copper and iron circuits, which by their cooling effects greatly assist the external surfaces. The surface which is exposed externally amounts to only

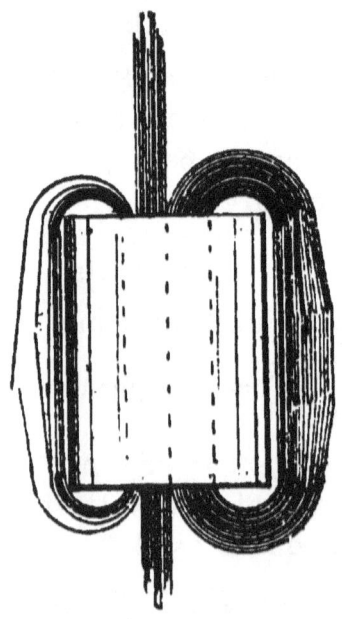

FIG. 21.

750 square inches, and hence the watts per square inch work out to $\frac{271}{750}$ = ·362 at full load, and to $\frac{151}{750}$ = ·201 at no load.

The method of construction of the iron circuit is explained in Fig. 21. In this transformer, as designed, the iron used consists of strips 3in. broad, arranged in three parallel bands, Fig. 20. This is done to facilitate the bending of the plates, and also to give increased ventilation through the iron. Also, it tends to reduce eddy currents in the iron if any appreciable magnetic leakage takes place out from the core through the broad plates. These strips of iron are cut to various lengths, as shown in Fig. 21, and laid out perfectly straight with alternate plates projecting each way. Then the formers containing the conductors can be slipped on over this rectangle of iron and secured in the centre of it. The next step is to bend the plates individually into their final positions. The reason of having them with alternate intervals at the ends is to secure that they break joint at the centre of the yoke. This keeps the magnetic resistance low, as explained in other designs. The weight and cost of this transformer will be found to compare favourably with those of the two previous designs, but it will not be correct to say from this that the type is the best. Until the best design of each type has been definitely fixed it is misleading to institute comparisons. So in this case the effects of less iron and more copper should be carefully investigated.

The three transformers worked out above by no means exhaust the types which have been used by the numerous manufacturers, but after carefully following the methods of design the reader should have no difficulty in working out details for any shape of core or of winding. Care must always be taken that in building up the core, no bolts, or other mechanical parts, form complete circuits embracing parts of the induction through the iron core. Otherwise secondary currents are induced, which add enormously to the losses.

The only remaining type which requires distinct treatment is the Hedgehog, an open-circuit transformer introduced by Messrs. Swinburne and Co. In this case the iron circuit is not closed, and the whole of the induction has to return from one end

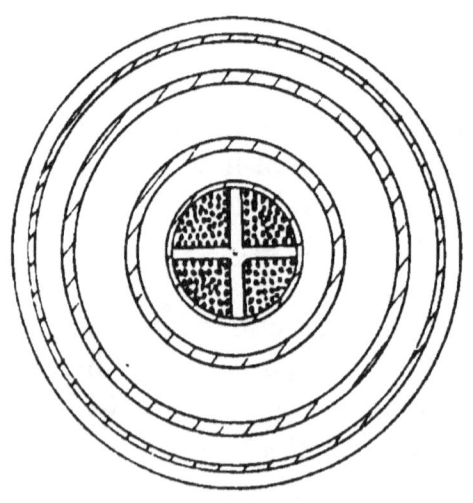

FIG. 22.

of the core to the other by the air. The advantage claimed for this transformer was that it had less iron loss than any ordinary closed-circuit transformer. This has been proved to be a fallacy by Dr. Fleming's measurements on a three and on a six kilowatt transformer, both of which had no-load losses as great as most closed-core types. He also showed that the manufacturers have been misled by a wattmeter which gave low power reading on inductive circuits. The grounds of the statement that there is less iron loss appears feasible at first sight. Suppose the transformer last designed was taken, Fig. 21, and the iron yokes were cut away just after the bend. This would reduce the weight of iron to one-half its former value. Then, if the same induction were used, the iron loss should be halved. But for the fact that the core is rectangular, we should now have a Hedgehog.

Figs. 22 and 23, shows the construction of this make of transformer. The core consists of thin iron wire stiffened by a gunmetal casting, which forms the support for all attachments. The iron wire is laid in between the flanges of this, and then bound tightly up with twine. Wooden flanges to support the copper wires are slipped on over the core. The winding of the coils is carried out much in the same way as described in other cases but for the fact that they are wound *in situ*, and that two sets of secondary winding are used, one each side of the primary. The ends of the wire core are spread out after the wooden flanges have been passed over them. The object of this spreading is to decrease the magnetic leakage along the core as much as possible.

The first abnormal point requiring notice is the large no-load current required. Thus, in the six-kilowatt tested by Dr. Fleming, the no-load current was 47·5 per cent. of the current required to furnish six kilowatts at the normal voltage of the primary. The actual loss at no load was 2·75 per cent., so

78 The Design of

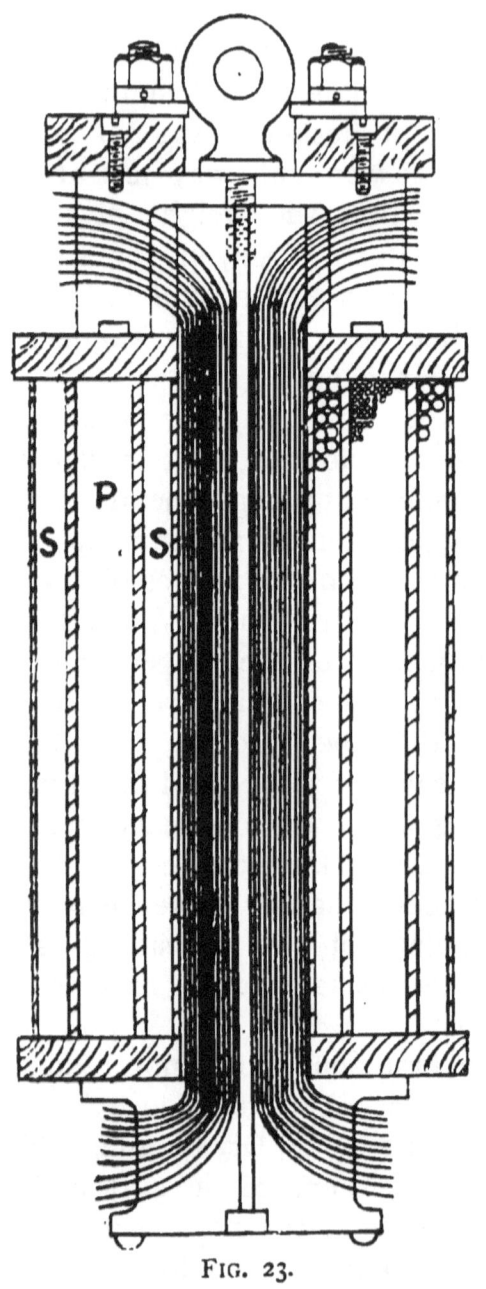

FIG. 23.

that the current required to magnetise the iron, i_μ, is $\sqrt{47\cdot5^2 - 2\cdot75^2} = 46\cdot6$ per cent., and

$$\frac{i_\mu}{i_1} = \frac{46\cdot6}{27\cdot5} = 17.$$

Now, in the closed-circuit transformers designed above, this ratio was slightly greater than unity, and in many transformers with superior brands of iron it falls below unity. The high value of i_μ in the Hedgehog transformer is due to the air resistance in the return path of the magnetic lines.

Again, it will be noticed that the iron loss was 2·75 per cent. of the total output, whereas if the transformer, Fig. 21, had its iron reduced by cutting away the yokes, the loss would be only 1·25 per cent. This, of course, may be explained by assuming that in the Hedgehog transformer the iron is worked at a much higher induction, but it is likely to be due to causes peculiar to the type. The lines of force in this transformer are supposed to flow along the full length of the iron and then to return by the air, and the turning back of the wire core at the ends does much to ensure this. Still, lines of force do leak out before the ends are reached to a much greater degree than occurs in any closed-circuit transformer. The result is, that to maintain the same E.M.F. in the wires, the total flux in the central sections has to be increased to make up for the smaller flux cutting the end conductors. Although the mean induction may be the same, the higher density in the centre will increase the total iron loss more than the diminution of density at the ends reduces it. Then, too, the leakage lines have to pass through the copper, and in so doing may cause eddy currents to an appreciable amount. These causes are likely to account for the no-load loss being considerably higher than would be expected from the quantity of iron used.

The large current at no load is the great disadvantage of this type, because although wattless in the transformer, it causes losses in all the conductors. Also, it heats the alternator armatures as much as if the apparent power were being actually supplied. Thus, if a station were equipped solely with Hedgehogs, at least 40 per cent. of the plant would have to be run continuously. The method suggested of overcoming this disadvantage is that of using condensers in parallel with the transformers to supply this large current. The dielectric losses in the condensers then have to be considered, and they are far from negligible. Even if the open circuit have the advantage claimed of little or no load loss, it does not follow that it is better than a closed circuit; for if two such transformers are taken and placed side by side, the iron at the ends would need a very slight increase of length in order to complete the iron circuit through the two. The result is a closed-circuit transformer of twice the output and about twice the iron loss. This loss will be reduced below the sum of the two individual transformer losses, owing to the uniformity of induction resulting from the completion of the iron circuit. The current at no load would of course be reduced to the normal amount required by a power factor of, say, 70 to 80 per cent.

Large transformers are expected to be more efficient than small ones, but still the combination thus made will show at once the advantage of having a closed core.

Magnetic Leakage of Transformers.—This branch of the subject is one that cannot be satisfactorily treated by calculation alone, and even when aided by previous experiments it is not easy to predict accurately the leakage effects in a new design. When a transformer is working a full load, the internal reactions cause fewer lines of force to pass through the secondary than through the primary, and hence we find a larger drop of volts than can be accounted for by loss in the copper. It is usual to deduct the known C R losses, and to call the remaining drop in pressure "magnetic leakage." When a transformer is working light the primary supplies the current necessary to magnetise the core and the secondary has no current in it, hence the action is simple and there is no differential magnetising effect. Now at full load in the third design the primary current has increased to 26·5 times its previous value, and we have the secondary current opposing the magnetising force of the primary. At no load, the magnetising force just inside the primary due to i_μ (·0864 ampere) would be $\dfrac{i_\mu \, \tau \, 1·76}{l}$, where τ = the turns per coil, and l = the length of the coil.

$$\therefore H = \frac{·0864 \times 865 \times 1·76}{23·5} = 5·6 \text{ in C.G.S. measure.}$$

So that the induction in the air would be very

small—*i.e.*, 5·6—compared to that in the iron. At full load the current rises to three amperes; so the magnetising force in the air space just inside the primary will now be

$$\frac{3 \times 865 \times 1\cdot 76}{23\cdot 5} = 195.$$

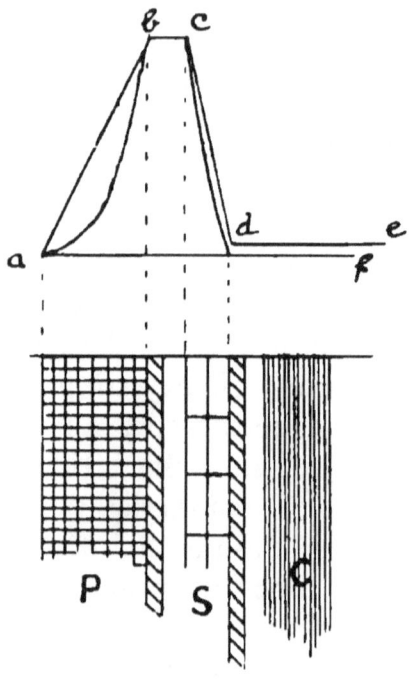

FIG. 24.

This will give an induction of 195 lines per centimetre, which, when multiplied by the area, gives a leakage flux which is appreciable.

The action is not confined to the actual space between the windings, but extends into the wire. Thus, in Fig. 24 the values of H are plotted over a section of the winding space. The magnetising force rises to its maximum value just inside the primary winding, and maintains this value till the secondary current causes it to fall. Halfway through the primary the force will be half the maximum, but as half the turns only are cut by the leakage lines generated here, the effect, if considered as acting on the whole number of turns in the primary, will be represented by half the ordinate at this point. Proceeding in this way, we get the total equivalent leakage effect, each point which is represented graphically by the curve lying under the lines, *a, b, c, d*. The area included between this curve and the abscissa, *a, f*, represents the total leakage flux when multiplied by the mean perimeter and constants depending on the scale to which the curve is drawn. This figure represents the number of C.G.S. lines which are acting on the primary only. It will be found that the flux thus obtained is a larger percentage of the total flux than the drop due to magnetic leakage is of the total voltage. This is because the leakage field thus summed up is in quadrature with the induction in the iron. Hence its effect must be worked out as acting at right angles to the flux in the iron. When this has been done, an approximate figure will be obtained generally under-estimating the leakage, because no notice has been taken of the return path of these leakage lines. With careful consideration of all these points, it is possible to obtain a fair approximation to the leakage effect, but when working on one type only experiments are more valuable. It is better in working out transformers to obtain the number of ampere-turns per given length of leakage path, and then by comparison with previous experiments to determine the probable leakage effects. When measuring the magnetic drop, great care must be taken that the circuit used for the load has neither induction nor capacity in it. Otherwise the results will be practically useless. Very little capacity will completely annul the drop and, on the other hand, induction increases it.

In connection with leakage it must be clearly remembered that it is only alteration in magnetic flux due to the secondary current that causes bad regulation. Leakage of the induction in the core across to the yokes, such as is found in all magnetic circuits, has no harmful effect on the regulation. Thus the Hedgehog transformer virtually has the return path of the magnetic lines as leakage, and yet it regulates as well or better than many closed circuits. This is due to the fact that the differences between the magnetising force of the primary at no load to its value at full load is nearly 3 to 1 instead of 26 to 1, as in the transformer above. This leakage, which does not affect the regulation, may, however, be very harmful in causing eddy currents in both iron and copper, and a Hedgehog transformer made with a core of iron plates would show this at once by giving a larger iron loss than with the wire core.

Alteration of Efficiency with Frequency.—The selection of the most economical frequency for alternate-current working from supply stations is influenced by so many different considerations that the present want of uniformity in this respect can hardly be a matter for surprise. It is most important that each company using the alternate-current system should arrange all their generating plant to work at the same number of cycles per second. If the prime movers are suitable, the alternators can then be run in parallel with each other, and even if the engine will not allow of parallel running, this uniformity in the frequency given by the different alternators in the same station is advantageous. Beyond this the matter has been largely left in the hands of the manufacturers. So we find that in America the high frequency of 133 periods per second has been extensively used because a large manufacturing company adopted it as their standard. In England, the practice is to use generally frequencies between 75 and 100, whereas on the Continent much lower values have been successfully adopted. It is out of the range of the present chapters to consider the whole grounds for and against either extreme in the adoption of a standard frequency. Roughly, the higher values tend to slightly reduce the first cost of the generating plant, and the use of motors is as yet only commercially possible where low frequencies are used. Whichever may be universally more economical, the fact remains that transformers have to be used, and it is the effect on the economy of the transformer that we have to consider.

From equation (1) the potential difference, as

measured by a Cardew voltmeter, at the terminals of a transformer on open circuit is

$$e = 4\cdot 45 \, F \, \tau \, n \, 10^{-8},$$

and F, the total flux, may be written $\mathfrak{B} a$ where $\mathfrak{B}$ equals the maximum induction, and a equals the cross-section of the iron core. Hence,

$$e = 4\cdot 45 \, \mathfrak{B} \, a \, \tau \, n \, 10^{-8}$$

With a transformer having a certain transforming ratio, say, 2,000/100 volts, the voltage, e, required on the secondary will be the same whatever the frequency of the alternating-current supply. This ensures that e shall be a constant quantity however the frequency may vary on different supply mains. We may therefore consider e as constant, and the quantities a and τ on the other side are also constants for any given transformer. Hence, if n changes, there must be an inverse change in $\mathfrak{B}$ to keep the E.M.F. up to the same value, or, in other words, $\mathfrak{B} n = \mathfrak{B}_1 n_1$, where $\mathfrak{B}_1 n_1$ are the new values of the induction and frequency respectively. Hence, in the transformers designed above for 100∼ the induction required would be doubled if the frequency were dropped to 50∼. This alteration in induction alters the value of the iron losses, and hence affects the efficiency of the transformer.

The subject of the loss in iron wire and plates when subjected to a magnetising force which varies has been a source of many investigations of late years. The conclusions arrived at by the many able experimenters on this subject are: first, that the hysteresis loss in the iron for any given induction is directly proportional to the number of reversals; second, from the many curves obtained it is found that the results are fairly represented by an equation containing $\mathfrak{B}$ to the 1·6th power. Hence the loss in hysteresis only may be expressed by the equation:

$$W_H = a\,n\,\mathfrak{B}^{1\cdot6} \quad \ldots \quad (5)$$

where W_H = the watts lost per pound of iron;

a is a constant depending on the iron used;

n = the frequency;

$\mathfrak{B}$ = the maximum induction.

The other part of the iron loss is that due to Foucault currents in the iron plates.

The determination of formulæ for calculation of these losses in plates requires some elementary calculus, and certain assumptions as to the path taken by the current have to be made. It will be sufficient to take the generally accepted formula for iron plate, which, when reduced to watts per pound of iron, becomes, roughly,

$$W_F = 5\,(t\,\mathfrak{B}\,n)^2\,10^{-9} \quad \ldots \quad (6)$$

at 100deg. F., where t = the thickness of the plate in inches and $\mathfrak{B}$ and n have values as above. Now, for any given transformer the square of the thickness of the plates may be merged into the constant, and the formula becomes

$$W_F = b\,(\mathfrak{B}\,n)^2,$$

where b is a constant. Hence the total loss is given by

$$W_H + W_F = a\,n\,\mathfrak{B}^{1\cdot6} + b\,(n\,\mathfrak{B})^2.$$

As n occurs in different powers in the two terms on the right-hand side of this equation, it will be readily seen that any curve drawn for the total iron losses for a given frequency cannot be used at any other frequency. It is, however, quite possible by careful experiments to verify the equation given above. This can best be done by taking power readings of the iron losses when the frequency is varied very considerably. Then the different set of results can be analysed by simple algebra. This can be more readily done with plates of greater thickness than those used in ordinary work, as in that case the expression for the eddy currents is a larger proportion of the total loss. The approximate truth of this formula and the corresponding expression for the eddy-current loss per pound of iron being established, we can proceed to divide the curve of iron losses, Fig. 4, into its two components. To do this, the loss in foucault currents in watts per pound for inductions varying from $\mathfrak{B} = 8,000$ to $\mathfrak{B} = 1,000$ must be calculated from (6), and the figures obtained when subtracted from the ordinates of curve, Fig. 4, leave the hysteresis loss per pound at 100 cycles per second. Thus at $\mathfrak{B} = 6,000$ for 10-mil plates we have

$$\begin{aligned}W_F &= 5\,(t\,\mathfrak{B}\,n)^2\,10^{-9} \\ &= 5 \times (\cdot 01 \times 6,000 \times 100)^2 \times 10^{-9} \\ &= 5 \times 6,000 \times 6,000 \times 10^{-9} \\ &= \cdot 180 \text{ watt per pound.}\end{aligned}$$

Now the ordinate on Curve 4 when $\mathfrak{B} = 6,000$ is 1·70. Therefore the hysteresis loss alone is 1·70 − ·18 = 1·52 watts per pound of iron.

Doing this for all the points, and plotting the results in a curve as before, we obtain Fig. 25, which gives the loss per pound of iron due to hysteresis only, at 100 ⌢∼. From this the losses at other frequencies can be obtained by simple proportion. It will be found on trial that the curve thus obtained is not quite in uniformity with the equation of $\mathfrak{B}$ to the 1·6th power. This is due to some qualities of the iron from which the original results were obtained. To explain fully the method of working with this curve it will be well to take the design No. 1 and to calculate the iron losses for that transformer when working at frequencies of 50, 75, 100, and 130 respectively.

We have from the table of results that for $n = 100$ $\mathfrak{B} = 3,580$ at the normal voltage; now $n\mathfrak{B} = n_1\mathfrak{B}_1$, so that the value of

$$\mathfrak{B}_1 \text{ for } n_1 = 50 \text{ is } \frac{3,580 \times 100}{50} = 7,160.$$

In the same way for $n_1 = 75$

$$\mathfrak{B}_1 = \frac{3,580 \times 100}{75} = 4,800$$

and at $n_1 = 130$, $\mathfrak{B}_1 = 2,770$.

We will in this case tabulate the results, as errors can then easily be seen and avoided. The fact that the hysteresis loss is directly proportional to the frequency now enables us to determine the total hysteresis loss in each of the above cases. Thus at $\mathfrak{B} = 7,160$ the loss per pound at 100 ⌢∼ is, from Fig. 25, 1·94 watts per pound. Therefore the loss at 50 ∿ in 172·5lb. will be

$$W_H = \frac{1·94 \times 172·5 \times 50}{100} = 168 \text{ watts.}$$

TABLE OF IRON LOSSES AND NO-LOAD CURRENTS FOR SHELL TYPE SIX-KILOWATT TRANSFORMER (DESIGN I) AT DIFFERENT FREQUENCIES.

Frequency.	50	75	100	130
$\mathfrak{B}$	7,160	4,800	3,580	2,770
Hysteresis watts	168	146	133·8	124
Foucault watts	11	11	11	11
Total loss	179	157	144·8	135
μ	2,650	2,300	1,960	1,680
i_μ	·127	·098	·0858	·0772
i_H	·089	·078	·0724	·0675
i	·155	·123	·112	·1025
Iron loss per cent.	2·98	2·62	2·42	2·25
i per cent.	5·16	4·16	3·72	3·42
Power factor	·57	·63	·648	·658

In the same way for $\mathfrak{B}$ = 4,800 and n = 75,

$$W_H = \frac{1·125 \times 172·5 \times 75}{100} = 146 \text{ watts.}$$

The rest of the values of W_H calculated in the same way are 133·8 for n = 100, and 124 watts for n = 130.

The Foucault loss is got from the formula (6)—*i.e.*, watts per pound = 5 $(t\, \mathfrak{B}\, n)^2\, 10^{-9}$.

In this transformer 10-mil plates were used, and the iron weighed 172·5lb.

Hence, $W_F = 172·5 \times 5 \times (·01 \times 7,160 \times 50)^2 \times 10^{-9}$.
     = 11 watts at 50 frequency.

But we saw above that $\mathfrak{B}\, n$ was always a constant if the voltage at the terminals of the transformer was not varied. So the eddy-current loss must also be a constant for all frequencies, as these two quantities are the only variables in the equation used above. Filling up the list, the total iron loss in each case can now be obtained by addition. It will be immediately observed that this increases rapidly with a decrease of the frequency. By halving the

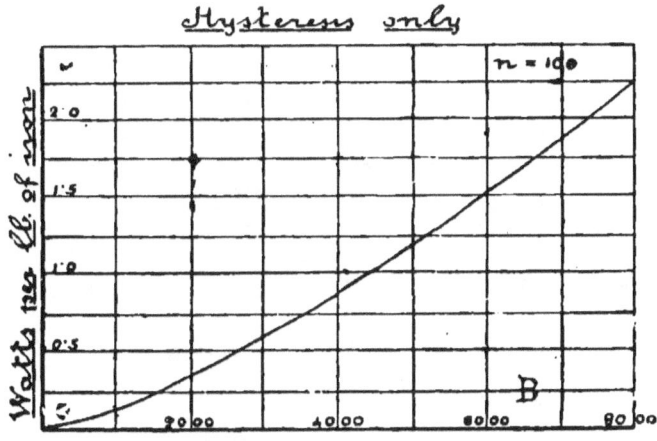

FIG. 25.

value of n, we have in this transformer increased the iron losses by some 25 per cent. Hence, if the transformer was designed to the heating limit at a frequency of 100, it would get too hot when placed on a circuit supplied at 50 frequency. This change in the iron loss is more marked in transformers using a higher induction than in the present design. To complete the list, we now require to find the exciting current at the different frequencies. The values of

μ for each induction can be at once taken from the curve Fig. 5. Then i_μ from (2) $= \dfrac{\mathfrak{B}}{\mu\, \tau_1 \times 1\cdot 76}$. As four values have to be calculated and part of the right-hand side, $\dfrac{l}{\tau_1 \times 1\cdot 76}$, is common to all cases, it may conveniently be worked out separately.

In this transformer, l, the mean length of the iron circuit was 123 cm. and $\tau_1 = 1{,}490$

therefore $\qquad \dfrac{l}{\tau_1 \times 1\cdot 76} = \cdot 0468$;

and $\qquad\qquad i_\mu = \dfrac{\mathfrak{B}}{\mu} \times \cdot 0468.$

Thus at $n = 50$, $\mathfrak{B} = 7{,}160$, and $\mu = 2{,}650$,

$$\therefore\ i_\mu = \frac{7{,}160}{2{,}650} \times \cdot 0468$$

$$= \cdot 127 \text{ ampere.}$$

The rest can be worked out in exactly the same way. The values of i_H are obtained by dividing the total iron loss by the voltage on the primary winding, which in this case was 2,000.

The magnetising current $i = \sqrt{i_H{}^2 + i_\mu{}^2}$, so in the first case $\qquad\qquad i = \sqrt{\cdot 089^2 \times \cdot 127^2}$
$\qquad\qquad\qquad\quad = \cdot 155$ ampere.

Completing the other values in the same way, it will then be better for reference to add two more lines, showing the iron loss and magnetising current respectively in percentages of the full-load values. The ratio of these two gives the power factor, which will be seen to fall with the frequency. This is not always so, but depends largely on the shape of the μ curve of the iron used.

The general conclusions from this investigation are briefly as follows : When the frequency at which the transformer is used is decreased, the loss in iron is increased, and there is consequently a fall in the efficiency of the transformer. (It must be remembered when considering the loss in the iron of a transformer, that although the loss may be a comparatively small percentage of the maximum output, it is a large percentage of the average load. Thus, while an increase of 25 per cent.—*i.e.*, from 2·42 to 2·98 per cent.—does not appear much, it may make the difference between profit and no profit in the central-station accounts.) Also the larger iron loss will increase the final temperature of the transformer. So, for low frequencies, more cooling surface should be provided. In other words, a larger transformer is required. This, of course, will raise the cost of the instrument for a given output. On the other hand, a transformer designed to have a good efficiency at a low frequency can be used more economically at higher frequencies, but the cost of manufacture will then be the prohibitive factor.

Quality of Iron.—This is the indeterminate quantity which causes uncertainty in the manufacture of transformers. Every batch of iron received should be kept separate, and the test of the transformers made from it then give a ready check on any falling off in the quality. The curves given (Figs. 4 and 5), which have been used in all the designs above, were taken from a transformer actually made. The iron was a better quality than the average brands of plate, but was by no means the best that can be obtained. With careful selection, iron can be found with which the losses per pound come out quite 30 per cent. below those given in Curve 4. It is in the permeability curve that the greater difference is noted with these superior qualities of iron. This curve, then, rises much more rapidly at low inductions, and the permeability throughout may be some 80 per cent. higher than the values given in Curve 5. The higher permeability reduces the current required to magnetise the iron, and consequently the power factor is increased. On referring to Dr. Fleming's list of tests, it will be seen that in one case he obtained a power factor as high as ·92. A special brand of mild steel is said to be used in this transformer, which must have a very high permeability to account for such a power factor.

Transformers for Different Methods of Supply.—The transformers used in lighting individual houses are very seldom worked at full load, and the average load is so small that the heating effect of the current in the copper circuit may be almost neglected. The transformer should, of course, be able to stand the full current for some hours without undue rise of temperature. The all-day efficiency is then fixed by the iron loss, and the copper might be reduced to save first cost if the reduction in section did not also affect the regulation. The sub-station system of supply is now coming into more general use. In this system the consumers are connected on to a low-tension network, which is fed at different places by transformers in sub-stations. In this way, even if the whole of the transformers are left continuously on the mains, a much higher load factor is obtained, owing to the different classes of buildings supplied. That is to say, the average load is a larger proportion of the maximum load, and hence the transformer is worked more economically. To still further increase the economy, arrangements are made for reducing the number of transformers in connection with the main as the load diminishes. So in these transformers and in those used in power transmission on a large scale, the watts wasted in the copper form a good proportion of the total loss. As the heat generated in the copper is liable to hurt the insulation, it is advisable in the large transformers used for central-station work to keep the percentage loss in the copper lower than in small transformers for separate installation work.

THE END.

"The Electrical Engineer"

WITH WHICH IS INCORPORATED

"ELECTRIC LIGHT."

Edited by C. H. W. BIGGS,

Assisted by Messrs. H. Swan, F. B. Lea, and A. Willbond.

"The Electrical Engineer" claims to be the most practical and the earliest-informed paper on all questions relating to Electric Lighting, and especially Municipal Lighting, as well as on all that relates to the Distribution and Application of Electric Power to all purposes. This paper was the first to make English readers acquainted with the three-phase system of distribution, the construction of incandescent lamps, the construction of alternate-current transformers, and has fully described and illustrated the various systems of practically laying mains. Every effort is made to keep readers fully acquainted with all progress in electrical matters.

It is the best technical paper for young electrical engineers.

It aims at and succeeds in giving early and accurate information about all that concerns the industry, and is absolutely free from any outside control.

It is an excellent advertising medium.

Every Friday.   Price Threepence.

139-140, SALISBURY COURT, FLEET STREET, E.C.

PRACTICAL ELECTRICAL ENGINEERING.—Being a complete treatise on the Construction and Management of Electrical Apparatus as used in Electric Lighting and the Electric Transmission of Power. 2 Vols. Imp. Quarto.

By VARIOUS AUTHORS.

With many Hundreds of Illustrations. PRICE £2 2s. 0d.

Among the information collected in and written specially for these volumes, are complete monographs by

Mr. GISBERT KAPP, M. Inst. C.E. on Dynamos, giving the principles on which the construction of these machines is founded, and a vast amount of new information as to the constructive details.

Mr. ANTHONY RECKENZAUN, M. Inst. E.E. considers the whole question of Electric traction on Railways and Tramways, describing the most recent practice and giving exhaustive details not only of the mechanism, but of practical considerations which are of the utmost commercial importance.

Mr. C. CAPITO, M. Inst. E.E. deals with steam—and considers the theory—as well as the practical construction of boilers and engines, more especially in connection with Electric lighting and transmission of power. This section alone contains no less than 371 illustrations.

Mr. HAMILTON KILGOUR considers carefully the theory of Electrical distribution upon which depends the right proportioning of mains in order to secure economical distribution. Other men have collected a complete description of the various systems practically applied, the materials used and methods of construction.

The above form but a portion of these volumes, but should prove of the utmost value to every one who professes to be an Electrical Engineer.

DYNAMOS, ALTERNATORS, AND TRANSFORMERS

BY

GISBERT KAPP, M.Inst.C.E., M.Inst.E.E.

ILLUSTRATED. PRICE 10s. 6d.

The book gives an exposition of the general principles underlying the construction of Dynamo-Electric Apparatus without the use of high mathematics and complicated methods of investigation, thus enabling the average engineering student and the average electrical engineer, even without previous knowledge, to easily follow the subject.

CHAPTER I.—Definition. Efficiency of Dynamo-Electric Apparatus. Measurement of Electric Energy. Principal Parts of Dynamo. Distinction between Dynamo and Alternator. Use and Power of these Machines.

CHAPTER II.—Scope of Theory. The Magnetic Field. Strength of Field. Units of Measurement. Physical and Mathematical Magnets. Field of a Mathematical Pole.

CHAPTER III.—Magnetic Moment. Measuring Weak Magnetic Fields. Attractive Force of Magnets. Practical Examples.

CHAPTER IV.—Action of Current upon Magnet. Field of a Current. Unit Current. Mechanical Force between Current and Magnet. Practical Examples. English System of Measurement.

CHAPTER V.—The Electromagnet. The Solenoid. Magnetic Permeability. Magnetic Force. Line Integral of Magnetic Force. Total Field. Practical Example. Extension of Theory to Solenoidal Electromagnets. Magnetic Resistance.

CHAPTER VI.—Magnetic Properties of Iron. Experimental Determination of Permeability. Hopkinson's Method. Energy of Magnetisation. Hysteresis.

CHAPTER VII.—Induced Electromotive Force. Cutting or Threading of Lines. Value of Induced Electromotive Force. C.G.S. Unit of Resistance. Fleming's Rule. Electromotive Force of Two-Pole Armature.

DYNAMOS *Continued.*

CHAPTER VIII.—Electromotive Force of Armature. Closed-Coil Armature. Winding. Bi-polar Winding. Multipolar Parallel Winding. Multipolar Series Winding. Multipolar Series and Parallel Winding.

CHAPTER IX.—Open-Coil Armatures. The Brush Armature. The Thomson-Houston Armature.

CHAPTER X.—Field Magnets. Two-Pole Fields. Multipolar Fields. Weight of Fields. Determination of Exciting Power. Predetermination of Characteristics.

CHAPTER XI.—Static and Dynamic Electromotive Force. Commutation of Current. Armature Back Ampere-Turns. Dynamic Characteristic. Armature Cross Ampere-Turns. Sparkless Collection.

CHAPTER XII.—Influence of Linear Dimensions on the Output. Very Small Dynamos. Critical Conditions. Large Dynamos. Limits of Output. Advantage of Multipolar Dynamos.

CHAPTER XIII.—Loss of Power in Dynamos. Eddy Currents in Pole-Pieces. Eddy Currents in External Conductors. Eddy Currents in the Armature Core. Eddy Currents in the Interior of Ring Armatures. Experimental Determination of Losses.

CHAPTER XIV.—Examples of Dynamos. Ronald Scott's Dynamo. Johnson and Phillips's Dynamo. Oerlikon Dynamo. Other Dynamos.

CHAPTER XV.—Elementary Alternator. Measurement of Electromotive Force. Fawcus and Cowan Dynamo. Electromotive Force of Alternators. Self-Induction in Armatures of Alternators. Clock Diagram. Power in Alternating-Current Circuit. Conditions for Maximum Power. Application to Motors.

CHAPTER XVI.—Working Conditions. Effect of Self-Induction. Effect of Capacity. Two Alternators Working on Same Circuit. Armature Reaction. Conditions of Stability. General Conclusions.

CHAPTER XVII.—Elementary Transformer. Shell and Core Type. Effect of Leakage. Open-Circuit Current. Working Diagrams.

CHAPTER XVIII.—Examples of Alternators. The Siemens Alternator. The Ferranti Alternator. Johnson and Phillips's Alternator. The Electric Construction Corporation's Alternator. The Gulcher Company's Alternator. The Mordey Alternator. The Kingdon Alternator.

ELECTRICAL DISTRIBUTION
ITS
THEORY AND PRACTICE.

PART I.—By MARTIN HAMILTON KILGOUR.
PART II.—By H. SWAN & C. H. W. BIGGS.

ILLUSTRATED. PRICE 10s. 6d.

PART I.

CHAPTER I.—Introduction.

CHAPTER II.—Constant Current (Series) System. Constant Current (Parallel) System. Feeders. Distributing Mains. Supply to Apparatus at a Variable Distance from the Generating Source.

CHAPTER III.—Economy in Design. Supply to Apparatus at a Constant Distance from the Generating Source.

CHAPTER IV.—Miscellaneous Problems on Feeders. Distributing Mains.

CHAPTER V.—Hypothesis. Calculations. Tables.

CHAPTER VI.—Examples. Tables of Square Roots. Particulars of Conductors.

PART II.

CHAPTER VII.—Bare Wire and Modification of Bare Wire. Crompton's System. Kennedy's System. St. James' and Pall Mall. Tomlinson's System.

CHAPTER VIII.—Practise in Paris. Compagnie d'Air Comprimé et d'Electricité. Continental Edison. Société du Secteur de la Place Clichy. Secteur des Champs Elysées. Halles Centrales.

CHAPTER IX.—Callender-Webber System. Callender Solid System.

CHAPTER X.—Brooks Oil Insulation System.

CHAPTER XI.—Ferranti Concentric Mains.

CHAPTER XII.—Modified Systems—Silvertown. St. Pancras Mains.

CHAPTER XIII.—Johnstone's Conduit System.

CHAPTER XIV.—Crompton and Chamen with Accumulators.

BY.

A. RECKENZAUN, M.I.E.E.

ILLUSTRATED. PRICE 10s. 6d.

This is the first serious attempt to consolidate and systematise the information of an important subject. Mr. Reckenzaun's experience is of the longest and widest, and this book deals not only with the scientific and practical problems met with in traction work, but enters somewhat into the financial aspect of the question.

CHAPTER I.—Early History. Magnetic Fields. Torque. Motor Efficiency. Insulation Resistance. Brake Tests. Electrical Horse-Power.

CHAPTER II.—Traction. High Pressure. Current, Stopping and Starting. Current Running. Energy Used.

CHAPTER III.—Advantages of Two Motors on Each Car. Calculations for Armatures. Constructive Details for Motors. Types of Tramcar Motors. Method of Suspension. Brushes and Brush-holders. Switches and Speed-Regulating Devices. Motor Trucks. Important Points in Tramcar Motors. Gearless Motors. Mechanical Transmission between Motor and Axle.

CHAPTER IV.—Description of Principal Systems of Overhead and Underground Construction used in Electric Traction. Poles. Trolleys and Trolley Wire. Overhead Curve Construction. Practical Hints on Overhead Construction. Insulators used in Electric Railway Work. Switchboards. Lightning Arresters and Cut-Outs.

CHAPTER V.—Secondary Batteries. Weight and Efficiency Deterioration.

CHAPTER VI. Portrush, Bessbrook-Newry, City and South London Railways. Central London Railway, Blackpool, Budapest, Leeds, and Halle Electric Tramways, West End Railway, Boston.

CHAPTER VII.—*Resume* of Writings and Expressions of Prominent Electrical Engineers on Electric Traction.

CHAPTER VIII.—Details of Working Expenses.

SECONDARY BATTERIES.

BY

J. T. NIBLETT.

ILLUSTRATED. PRICE 3s. 6d.

This little treatise is an attempt to bring together and describe the recent commercial developments of Planté's original discovery, the Lead Secondary Battery.

CHAPTER I.—Introductory.

CHAPTER II.—Cells of Planté type such as those of De Meritens, De Kabath, Reynier, Montaud, Elwell-Parker, Cheswright, Epstein, Woodward, Dujardin, Dujardin-Drake-Gorham, Crompton-Howell.

CHAPTER III.—Pasted or Faure Storage Cells, such as: E.P.S., Eickemeyer, Gadot, Hagen, Jacquet, Pitkin-Holden, Ernst, Payen, Carpenter, Bailey, Knowles, Laurent-Cély, Hering, Tudor, Pollak, Gibson, Frankland, Currie, Pumpelly, Hatch, Winkler, Tommasi, Bristöl, Lithanode.

CHAPTER IV.—Lead-Zinc, Copper, Alkaline, and other batteries such as Reynier, Bailly, Hedges, Lalande-Chaperon, Thomson-Houston, Entz-Phillips, Main, Barker, Kalischer, Marx, Taplow, Osbo, Jablochkoff.

CHAPTER V.—The Electrolyte, etc.—Solid Storage Cells, such as, Barber-Starkey, Schoop, Parker, Volk, Crova and Garbe, Roux.—Spray.

CHAPTER VI.—Appendix—Table for the conversion of Measures, Thermometric Scales, Lead and its Impurities, Plumbic Hydrates, Sulphuric Acid, English Oil of Vitrol, Electrical Units, Ohm's Law, Measuring the Internal Resistance of Voltaic Cells, Capacity and Efficiency of Storage Cells, etc.

PORTATIVE ELECTRICITY.

BY

J. T. NIBLETT.

ILLUSTRATED. PRICE 2s. 6d.

This is a treatise on the Application, Method of Construction, and the Management of Portable Secondary Batteries.

INTRODUCTION.—A brief résumé of Electrical Discovery.

PART I.—Portative Electricity in Mining Operations, describing some of the Lamps used, and Instruments for the Detection of Dangerous Gases.

Safety Electric Hand-lamps for Domestic Purposes, for Customs' Officers, for Meter Inspectors, and for Firemen.

Portative Electricity for Domestic Purposes, for Reading, for Driving Motors, for Sterilising Water.

Application to Medical and Scientific Purposes.

Uses for Military Purposes, for Land Work, At Sea.

Uses for Lighting Vehicles, Railway Carriages, Omnibuses, Tramcars, Broughams, Cycles.

Traction by means of Portative Electricity.

Applications to Decorative Purposes, for Dinner Tables, Shop Windows, Fountains, Jewellery, on the Stage, Personal Adornment.

PART II.—Secondary Cells, Planté's, Crompton-Howell, Epstein, D. P., Faure, E. P. S., Lithanode, Pitkin, Bristöl. Types of Secondary Cell. Niblett's Solid Cell.

PART III.—The Management of Portable Apparatus for storing Electrical Energy. Methods of developing Electrical Energy. Thermopiles. Primary Batteries. Charging Instructions.

The Charging Instructions are very complete, and should be of immense service to those who use small Portable Cells.

POPULAR ELECTRIC LIGHTING.

BY

CAPTAIN E. IRONSIDE BAX.

ILLUSTRATED. PRICE 2s.

CHAPTER I.—Advantages of the Electric Light.
CHAPTER II.—Source of Supply. High and Low Tension. Continuous and Alternating Current. Storage Batteries. Electromotive Force. Volt. Transformers. Distance of Generating Station from Consumer's House. Number of Stations Worked by Company.
CHAPTER III.—Street Mains. Branches Laid for Consumers. Consumers Living in Flats. Consumers beyond Twenty Yards from a Main. Cost of Mains in Private Grounds.
CHAPTER IV.—Number and Candle-Power of Lamps Required for Rooms. Incandescent Lamps. Notification of Additional Lamps to be given to Supply Company. Wiring of Houses. Cheap Wiring to be Avoided. Insulating Material. Leakage. Switches. Economy Effected by Arrangement of Switches. Arc Lamps.
CHAPTER V.—Safety Fuses. Wiring to Pass Test. Connections of House and Street Mains. Switch-board. Fittings. Short Circuit. Defects. Incandescent Lamps. Pressure. Voltage. Life of Lamps. Breakage of Lamps. Extinctions. Spare Fuses. Caution as to Substituting Wire for Fuses. What to do in case of Fire. Fluctuations. True Test of Efficiency of Supply. Electricity v. Gas in Fogs.
CHAPTER VI.—Supply of Energy by Contract. Electricity Meters. Examination of Meters by Board of Trade. Different Types of Meter. How to Read Meter. Board of Trade Unit. The Ampère. Reliability of Meters.
CHAPTER VII.—Hydraulic Analogy. The Watt. Reading of Meter taken by Supply Company Current may be cut off if not paid for. How Consumer may make his own Test as to Accuracy of Meter. Provision made by Act in case of Dispute. Course to be Adopted by Consumer in case of Dispute. Relative cost of Electricity and Gas. Electric Light and Eyesight.
CHAPTER VIII.—Estimates for Wiring Houses of Various Sizes, Supplying Fittings and Estimated Approximate Yearly Cost of Current. Cost of Arc Lamp Lighting.
CHAPTER IX.—Private Installations. Machinery and Plant Necessary. Estimates Showing Cost for Houses of Different Sizes.
CHAPTER X.—Electric Motors. Heating and Cooking by Electricity.

ECONOMICS OF IRON AND STEEL.

BY

H. J. SKELTON.

ILLUSTRATED. PRICE 5s.

CHAPTER I.—Ironstone and Iron Ore.

CHAPTER II.—Pig Irons.

CHAPTER III.—Pig Iron Warrants.

CHAPTER IV.—Iron Castings.

CHAPTER V.—On Testing Cast Iron.

CHAPTER VI.—Puddled Iron.

CHAPTER VII.—Staffordshire and other Irons.

CHAPTER VIII.—Sheet and Plain Iron.

CHAPTER IX.—Scrap and other Qualities of Iron.

CHAPTER X.—Galvanised Sheet Iron.

CHAPTER XI.—On Testing Wrought Iron and the Tests for the Same.

CHAPTER XII.—On Testing Wrought Iron in the Smithy.

CHAPTER XIII.—Steel.

CHAPTER XIV.—On the Tests Applied to Ascertain the Quality of Steel.

CHAPTER XV.—Tin Plates.

CHAPTER XVI.—Quarterly Meetings of the Iron Trade.

CHAPTER XVII.—General Considerations Affecting the Purchase of Iron and Steel.

CHAPTER XVIII.—Sizes and Sections of Rolled Iron and Steel, Including Memoranda and Tables of Extras.

FIRST PRINCIPLES OF ELECTRICAL ENGINEERING.

BY

C. H. W. BIGGS.

Editor of "The Electrical Engineer" & "The Contract Journal."

Second Edition. Crown 8vo. Price 2s. 6d.

THE first edition of this book has long been out of print, but the author has been unable owing to pressure of other work to arrange for the publication of the new edition. The original book was, as was expected, severely handled by the critics. As a matter of fact it was intended to be a bone of contention, and the author has found no reason to withdraw from the position he then took up. Some portions of the book have been rearranged, and in one or two cases where condensation led to the views promulgated being mistaken, the matter has been expanded and more fully illustrated. The book is intended to be a first book for electrical engineers and avoids as much as possible the discussion of electrical questions that have no bearing upon the probable future work of the reader.

CHAPTER I.—Introductory.

CHAPTER II.—Deals with the Conductive Circuit, the Inductive Circuit, and the Magnetic Circuit.

CHAPTER III.—Discusses the Production of Electrical Pressure or Difference of Pressure. The Use of Electro-Graphics Loops of Force, and Interaction among Circuits.

CHAPTER IV.—Kinds of Dynamos. Characteristic Curves. Self-induction. Motors. Distribution. Simple Measurements. How to Measure Current. The Tangent Galvanometer. Ampere-meters and Volt-meters. Cardew Voltmeter. Wheatstone Bridge Method.

The new edition is printed in similar type to the other technical works issued by the publishers, as it is thought the larger type used is more suitable for readers.

First Principles of Mechanical Engineering,

BY

JOHN IMRAY, WITH ADDITIONS BY C. H. W. BIGGS.

ILLUSTRATED. PRICE 3s 6d

An attempt is made in this book to explain the principles of Mechanical Engineering in simple language, and without the aid of abstruse mathematics. The calculations are generally arithmetical, and such as come well within the comprehension of the beginner.

CHAPTER I.—Mechanics of Antiquity. Modern Machinery Mechanics and Chemistry. Statical and Dynamical Machines. Strength of Materials. Nomenclature. Unit of Work. Source of Energy. Application of Energy. Friction. Governing. Nature of Machines. Mechanical Education.

CHAPTER II.—Mechanical Drawing. Use of Drawing. Plane Surfaces. Drawing Cubes. Drawing Cylinders. Projection. Sections. Perspective, etc. Instruments. Scale of Drawings.

CHAPTER III.—Strength of Materials. Strains and Stresses. Tension. Compression. Transverse Stress. Deflection and Disposal of Materials. Calculations. Torsion. Shafts. Clipping and Shearing Stress.

CHAPTER IV.—Sources of Mechanical Power. Muscular Force. Wind. Windmills. Water. Waterwheels. Turbines. Weight. Falling Bodies. Springs.

CHAPTER V.—Heat. Expansive Power of Heat. Elasticity of Gases. Temperature and Pressure. Condensation. Expansion. Steam Boilers. Steam Engines.

CHAPTER VI.—Electricity. Chemical Action.

CHAPTER VII.—Transmission of Power. Rotary Motion. Couplings. Clutches. Plummer Blocks. Pulleys or Drums. Toothed Wheels. Reciprocating Motion. Discontinuous Motion. Ratchet. Cam. Mangle Motion. Reversing Gear. Dynamometer. Brakes etc.

CHAPTER VIII.—Transmission of Power. Water. Compressed Air. Electricity.

Theory and Practice of Electro-Deposition, including every known mode of depositing metals, preparing metals for immersion, taking moulds and rendering them conducting. By Dr. G. Gore. F.R.S. Illustrated. Crown 8vo. Price 1s. 6d.

This little work is too well known to need a lengthy description. It has long been acknowledged to be the simplest elementary manual on the subject, and Dr. Gore's ability and great care to put his matter clearly, is nowhere better shown than in this little book.

EDITED BY C. H. W. BIGGS.

The Electrical Engineer.—A journal of Electrical Engineering. Published every Friday. Price 3d. Contains more, and earlier information about practical work than any other similar technical paper.

EDITED BY C. H. W. BIGGS.

The Contract Journal.—Published every Wednesday, Price 6d., has devoted its attention during its fourteen years of existence to the discussion of all questions relating to contracts. It has full reports of all matters concerning municipal contracts, and has gained a position unique with regard to technical journals, and is without doubt considered the authority in the great subject of contracts and tenders.

SHORTLY.

First Principles of Building. By A. Black., C.E. Illustrated.

THE MARINE ENGINEERS ELECTRICAL POCKET BOOK.

By M. Sutherland (Electrical Engineer at W. Denny & Bros. Shipbuilding Yard, Dumbarton.)

The Marine Engineers Electrical Pocket Book.

PREFACE.

The writer has frequently been asked by marine engineers, to recommend a book which would give them a sufficient insight into electrical engineering to enable them to understand ship installations without going too deeply into electrical matters in general, and though there are many excellent books on practical electrical engineering, he has not been able to find one which exactly fulfils these conditions. His object therefore, has been to produce a compact and handy volume, which should give an insight into magnetism and electricity, with descriptions of the various systems of wiring, generating plant, fittings, measuring instruments etc., used on board ship, avoiding as much as possible, all matter not directly applicable to work of this nature.

ALTERNATE CURRENT TRANSFORMER DESIGN. By R. W. Weekes.

Whit. Sch. A.M.I.C.E. Crown 8vo. Price 2s.

This book is one of a new series, intended to show engineers and manufacturers the exact method of using our acquired knowledge in the design and construction of apparatus. Mr. Weekes has taken a number of different types and calculated out fully the dimensions of the various parts, showing each step in the calculation. Diagrams are given drawn to scale, and a summary of sizes, weights, losses and costs given at the end of each design.

At the present time transformers play an important part in high pressure distribution, and it is of the greatest importance that they be constructed to give as little loss as possible.

ELECTRIC LIGHT AND POWER.

By Arthur F. Guy. A.M.I.E.E.

This book is written from an intensely practical point of view, and deals with the subject as known from experience. The following description of the contents of the first part of the book will give a very good idea of the author's views.

> Evolution of Electrical Engineering showing the use of Electric lighting. The action of parliament. The advantages of the electric light.

> Motive power. Artificial. Conservation of Energy. Coal as fuel. Work and horse power. Mechanical equivalent of heat. Gas and oil engines. Water power. Ohm's law and the electric circuit. Heating effects of the circuit. The dynamo. The magnetic circuit. Generation of Currents. Alternating currents. Field magnet winding. Armature winding. Working in parallel. Notes on running, etc.

The Dynamo.—By C. C. Hawkins, A. Inst. E. E., and F. Wallis, A.Inst.E.E. With numerous Illustrations. Price 10/6.

The Management of Accumulators.—By Sir D. Salomons. Price 5s. (Forming the first part of the Seventh revised and enlarged Edition of the same Author's work, entitled 'Electric Light Installations and the Management of Accumulators.')

Lightning Conductors and Lightning Guards.—By Oliver J. Lodge, LL.D., D.Sc., F.R.S., M.I.E.E. In one volume, with numerous Illustrations. Crown 8vo. Price 15s.

Electric-Light Cables, and the Distribution of Electricity.—By Stuart A. Russell, A.M.Inst.C.E., M.I.E.E. With 107 Illustrations. Price 7s.6d.

Alternating Currents of Electricity.—By Thomas H. Blakesley, M.A., M.Inst.C.E. Hon. Sec. of the Physical Society. Third Edition, enlarged. Price 5s.

Electric Transmission of Energy, and its Transformation, Subdivision, and Distribution.—By Gisbert Kapp, M.Inst.C.E., M.I.E.E. A practical handbook, with numerous Illustrations. Third Edition, thoroughly revised and enlarged. Crown 8vo. Price 7s.6d.

The Telephone.—By W. H. Preece, F.R.S., and Julius Maier, Ph.D. With 290 Illustrations, Appendix, Tables, and full Index. Price 12s. 6d.

Maycock's First Book of Electricity and Magnetism. 84 Illustrations. Price 2s. 6d.

Electricity in our Homes and Workshops.—By Sydney F. Walker, M.I.E.E., A.M.Inst.C.E. A Practical Treatise on Auxiliary Electrical Apparatus. With numerous Illustrations. Second Edition. Price 5s.

Electrical Instrument-makers for Amateurs.—By S R. Bottone. A Practical Handbook. With 71 Illustrations. Fifth Edition, revised and enlarged. Price 3s.

www.ingramcontent.com/pod-product-compliance
Lightning Source LLC
Chambersburg PA
CBHW021944160426
43195CB00011B/1216